# Wilderness and the Changing American West

Gundars Rudzitis

*University of Idaho*

**John Wiley & Sons, Inc.**

New York • Chichester • Brisbane • Toronto • Singapore • Weinheim

This text is printed on acid-free paper.



*Library of Congress Cataloging in Publication Data:*

Rudzitis, Gundars, 1943—
Wilderness and the Changing American West / Gundars Rudzitis.
p. cm.
ISBN 0-471-13396-5 (pbk. : alk. paper)
1. Wilderness areas—West (U.S.)—Management. 2. West (U.S.)
I. Title.
QH76.5.W34R84 1996
333.78m'2'0978—dc20 96–15426

Printed in the United States of America

10 9 8 7 6 5 4 3 2 1

# Wilderness and the Changing American West

*For Rosemary Ann Streatfeild,*
*Kristine, and Erik*

# Contents

# Preface

Twenty years ago, armed with a Ph.D., I drove from Chicago to Austin, which I thought was the heart of the American West, to teach at the University of Texas. There I found people who wore cowboy boots, but the male students were more likely to drive a Buick Regal than a Ford pickup. The female students were more likely to wear high heels and perfume than Levis. The typical Texan student came not from a ranch, but from suburban Dallas or Houston. I had arrived in the urban West, where the influence of mythology still prevailed. The school mascot was a longhorn steer, but an oil well would have been more symbolic of the "real" Texas. And although open spaces were plentiful, they were private, fenced-off spaces. Public lands were scarce.

I was in the heart of the privatized modern West. It was a New West, to which people were flocking from other parts of the country. There were many instant Texans, as people would declare by wearing caps that said they were "Naturalized Texans." Being Western and Texan was critical to one's identity. Remnants of an older West remained even as people wore pinstripe suits, cowboy hats, and boots to conduct their everyday business.

I arrived at a much different West when I moved to Idaho, the state that outsiders think of as the home of the famous potatoes, but that those who live here experience as a living reflection of the Old West. Students wore boots (cowboy, logging, or hiking) and jeans, and drove pickups. Most were from rural, not urban, areas. Attendance dropped every year when hunting season opened. In Texas, as expected, controversies revolved around keeping energy prices high and the consequences of rapid urban-suburban growth. In Idaho and throughout the Old West, the issues were how public lands were managed and what restrictions were either inappropriate, unjustified, or just plain idiotic. The presence of public lands and the agencies that managed them dominated the debates at local and state levels.

The Old West also appeared to be under attack from both within and without. An Old West rooted in conquest and the extraction of resources was becoming more and more of an anachronism, because a New West based on preserving the "wildness" of the Old West was challenging the destruction of that "wildness." An American West born out of wilderness needed to have federal legislation enacted both to protect wilderness and to keep it from disappearing.

Once protected, the wilderness had to be managed. Ironically, it has been managed by the very agencies that initially did not want it designated as a protected resource. Despite the sometimes best efforts of the managing agencies, the old management system rooted in the West of the past has not worked well. Until recently, most Americans have not been well informed or have trusted too much those who manage our national wildlands. Reform is overdue, and simply updating old proposals for coordination or agency restructuring will not work. I discuss some of the more innovative approaches to reform and present my own proposals as well. Some approaches, such as those to give the federal wildlands to the states or to privatize them in the name of efficiency, I find both intellectually lacking and potentially destructive of the fabric and soul of the American West.

The attitudes toward our wilderness lands and their value to us has changed as we have matured as a nation. In the past, public wildlands have been managed without consulting the American public. This is apparent in the increasing number of conflicts about how these wildlands should be used that are being reported by the national media. Decisions that used to be made quietly—whether they were about preserving wildness for spotted owls, opening up wilderness areas in Alaska to oil drilling, or making wilderness more "wild" by reintroducing grizzly bears and wolves to appropriate habitats—have become part of a national debate.

People throughout the United States and in communities in the West are becoming more supportive of keeping our federal lands wild. New people have been moving into the American West in record numbers, and they have changed the economies of their adopted communities. This phenomenon has been called the New West, and how and why it has come about is only beginning to receive the serious attention that it merits. One reaction, though clearly a minority one, is the rise of movements and even hate groups who see the Old West as a refuge where they can hide.

I try to discuss what all these changes, reactions, and counterreactions mean for the American West, the people who live there, other Americans, and people outside the United States for whom the region represents the promise of what our country can be. To my dismay, much of the discussion of federal land management issues or of why the West is changing ignores the original settlers, the Native Americans. I have tried, however imperfectly, to provide some possible ways for us not to ignore our predecessors, but to learn from them. How we choose to live with our wilderness in the future will say a lot about the direction our society takes, and there are forks in the road immediately ahead that will require us to make important decisions.

To me, the American West is wilderness, yet wilderness and the wild mean different things to different people. There is a capitalized, majestic Wilderness and a wilderness that exists on a smaller scale, but is no less valuable. I begin with the image of a raw wilderness that presented itself to the Europeans who first encountered it, and discuss the designation of a federally specified class of wilderness. Despite this designation of an "official" wilderness, it is an artificial distinction to argue that the other federal lands, however heavily used to produce commodities are not, or cannot be, part of the greater public lands wilderness. Fortunately, though wildness of these lands may be diminished and abused, it has not totally disappeared. So, I hope to be forgiven for moving back and forth in referring to both "official" designated wilderness and the rest of the federal estates as our wildlands.

The American West and its public wildlands is a region both unique and easy to love. Though I concentrate on the Northwest and the Northern Rockies because those are wildlands through which I have trod the most and have become increasingly concerned about, the issues and the conclusions drawn should apply to other parts of the West as well. Fortunately, the West is full of people who, though they may disagree about many things, agree that federal wildlands are perhaps the defining feature that gives their lives meaning and characterizes a Western culture past, present, and future.

# Acknowledgments

I wanted to write this book soon after I moved to Idaho and became interested in public lands and especially wilderness. Growing up in the East and living and traveling mainly in the Midwest and South, I was more accustomed to environmental issues of air and water pollution and hazardous substances. Upon my arrival at the University of Idaho, I was fortunate to meet and subsequently become friends with Michael Frome, who was at the time Conservation Writer in Residence at the College of Forestry. Michael both inspired and encouraged me to write about issues relevant to the West. He was also an example of someone who is not afraid to take a stand after looking at the facts. I have tried, however imperfectly, to do the same and have avoided the traditional academic approach of playing it safe. I have followed my interpretation of the "facts" to where I think they should lead me, knowing that others may disagree with my conclusions and recommendations. That is how it should be.

It doesn't seem that long ago that I was having a beer with fellow geographer John Alwin while talking about how both of us wanted to write books that reached out to the general public as well as to traditional academic and public policy audiences. He has already done so with books about Washington and Montana. As I was talking about the need to read and to do more research about wilderness, John said something to the effect, "Come on, you could just sit down and write this book." I was taken aback but realized that he was right. Later, foolishly I announced in my fall 1994 seminar that I would have the book done before the end of the semester. It has taken longer and would have taken even longer without the help of various people, some of whom I would like to thank here.

I wish to thank my students, who in many ways were my instructors as I started writing this book and tried out ideas and asked many questions along the way. They also suffered at times from ill-informed ideas and responded with many wise insights and personal experiences. Perhaps because of the nature of geography and the subject matter, my seminars and classes attracted people from a wide variety of disciplines in the social and natural sciences, as well as the humanities, which led to wide-ranging discussions. In particular, among those who enlivened my teaching and learning, I would like to thank Marty Anderson, Scott Birkey, Anne Black, Bill Carlson, Jason Doolittle, Jack Erb, Karen Feary, David Fosdeck, Steven Gill, Mark Haugen, Karen Kaasik Dean, Lance Krull, Erik Kummert, Eversley "Teddy" Linley, James Mackey, Shaun Maxey, Russ McCabe, Bill Owens, Ken Preston, Christina Sanders, Philip Smith, Cynthia Tauber, James "Smokey" Thompson, Chris Wall, and Courtney Watson.

I benefited greatly from discussions with a wide variety of people who are concerned with the future of the West. I especially thank Bob Greene for having the best bookstore in the "Real West." He has guided me to many a good book, and his friendship, conversation, and wit, and those of his staff and patrons have made Bookpeople an institution in Moscow, Idaho, without which the town just would not be the same.

With apologies to those whom I have inadvertently left out or who wish to remain anonymous, hearty thanks to Mike Anderson, Sue Armitage, Dennis Baird, Roger Bolton, Sam Couch, Dale Crane, Bob Dale, Don Dahman, Tim Eaton, Cindy Fisher, Jim Fisher, Inese Gruber, Frank Gruber, Joel Hamilton, Bob Hautala, Paul Hirt, Harley Johansen, Jerry Johnson, Tom Kovalicky, Paul Lindholt, Dick Morrill, Jon Miller, Scott Morris, John Norton, Tom Power, Gunita Pujate, Ray Rasker, Karel Stozek, Bill Swagerty, Christiane von Reichert, Paul von Reichert, George Tolley, Tom Lamar, Nancy Taylor and the rest of the gang at the Palouse Clearwater Environmental Institute, Jim Reece, Al Rouyer and the infamous DeMoura running team. I especially benefited from the comments of Olen Paul Matthews on the entire manuscript and Mike Scott on the ecosystem chapter. My colleagues in the Geography Department, where humor is the rule of the day, provided a relaxed atmosphere.

This past year I have been fortunate to have John Hintz and Christy Watrous as research assistants; they have provided invaluable assistance, some of which is reflected in these pages. They have made the writing of several chapters much easier and more pleasurable. John Hintz also did an outstanding job of providing the maps for the book in a timely fashion. I also benefited greatly from the support of grants from the National Science Foundation.

Judith Scott took my individual chapter files and put them all together in a file that was useful to my editors. Dean Harshbarger solicited my book proposal. My editor at John Wiley & Sons, Philip Manor, took my original proposal and got it approved in what must be record time, and he has prodded and coaxed me along the way to keep me near deadlines. He kept reminding me that the West is indeed changing rapidly and the sooner I documented some of these changes and their implications, the better. I hope the book proves him right. The production process was expertly guided by Donna Conte. Warren Freeman corrected and sharpened my prose without changing the tone or content, to the ultimate benefit of the reader.

Despite all this wonderful help, I am fully responsible for any errors of fact or interpretation. My apologies also to those whose advice I did not take and to some who tried to convince me gently and otherwise of the folly of my ways. I believe strongly in the need to keep the landscape of the West wild but respect the rights of others to disagree about just what that means. Let them write their own books, or, more important, participate democratically in the public debate to which I hope this book contributes. The West is a region that is easy to romanticize and love, and there is nothing wrong with romance or love. We need more of it in our public discourse.

Finally, without the indulgence and support of Rosemary Streatfeild and my children, Kristine and Erik, writing this book would have been less fun. They supported me in a variety of ways that is difficult to put into words. Rosemary got me out into the wild when I needed to be there, though often behind her on the trail.

She kept me on track when I threatened to falter, procrastinate, talk, and not write. She also did frontline editing and had the uncanny ability to know when I had written something after midnight.

I wrote much of this book on laptop computer. One day I overheard my daughter telling a friend, "you can always find him on the couch writing on his laptop." One reason I wrote this book was in the hope that our public lands would remain wild for Kristine and Erik, their generation, and others who follow seven generations and beyond. Writing about wilderness is one thing; spending time with loved ones, family, and friends in the wilderness is another. People may prefer different kinds of wilderness, but diminishing our choices can only do us all harm.

*Gundars Rudzitis*
Moscow, Idaho

# Wilderness and the Changing American West

## *Chapter One*

# Wilderness and the American West

The controversy over wilderness lands of the West can provide a glimpse into the future of the United States. In the American West, more so than in other regions of the country, that controversy has a direct impact on the landscape, much of which is not a private landscape, but a national one that belongs to all the citizens of the country. What happens to the Western landscape reflects how the mythical American past is forced to confront a present-day reality and to form the outlines of a future American West.

A conversation in a cafe-bar in a small town in Northeast Oregon expresses typical complaints about how and why changes are taking place on the Western landscape. Mike, an employee of a large corporation, is complaining that he was given two weeks' notice to relocate to California or Georgia or lose his job. "I told them I wouldn't go, and so I sold my house and moved down here. What gets me is the company lied to us for three years about not moving and then just like that you get two weeks to get up and leave. It is good old-fashioned greed. The company doesn't give a damn about us and our families."

Dan, the owner of a local logging company, sympathizes with Mike. "It used to be you could make a good living out here. Now with all the restrictions on logging or, worse, not allowing us to log on large parts of our public lands . . . making a living is getting harder and harder." For people like Mike and Dan, pursuing other types of jobs is undesirable because it would mean giving up a way of life they would find hard to duplicate outside the inner West.

Conversations such as these can be heard all over small towns in the West. They make clear that private, corporate, and federal government decisions continue to have a major impact on the landscape of the West. This uncertainty is nothing new. Boom and bust cycles, ghost towns, and controversies are an integral part of the history of the region.

What is different about the hostility toward government management and ownership of Western lands is that it is being more and more openly expressed. Often an implied threat lurks beneath the surface: "Save the West or else." Fear of job loss and change drives the local hostility toward the federal government. Saving the West really means "save our jobs." The hostility is directed at the federal government because most of the lands are in the public domain. The frustration felt by the people is emphasized by journalists and the media as well as by politicians hoping to use it for their own purposes. And yet the anger is often misdirected, because it ignores the inevitable changes or places the old West in a mythical context.

There are federally designated wilderness areas in many states, but most of them are in the American West. Originally these were the lands that could not be used or developed commercially, and therefore were set aside to be conserved or managed for people mainly "back East."[1] There were fears that, if public lands were not set aside, the private timber companies would cut and run, creating a vast denuded landscape. Such overcutting would create perpetual timber shortages in the United States. To prevent this, a professional government cadre of forest managers was established.

The timber shortage argument echoes through time right up to the present. In the 1990s, overcutting on public lands, technological change, and increased productivity—combined with past neglect of environmental considerations and changes in public awareness and attitudes—have all contributed to making less timber available from public lands and temporarily driving up prices. A cyclical pattern for timber and other resources in the region is an ongoing phenomenon. This is all part of a changing American West. Some of the changes are quite contentious, because they are forcing a reevaluation of many of the myths that have been accepted as defining not just the West, but the United States as well.

## WHERE IS THE WEST?

What is the American West, and why do we hear about the Old and the New or Changing West? To a nation raised on stories about pioneers, cowboys, and Indians, the American West is both a place and a part of our history. It is a more exciting history than that of the East, with its stereotypically elegant New Englanders, or of the South, imbued with the remnants of slavery, plantations, and racism. Contrary to some of the negative images in other regions, the American West represents space, freedom, individuality, and conquest—qualities idealized in movies and television programs.

Just where is the American West? When people are asked to define it, a wide variety of responses are evoked, ranging from climatic and other descriptions of the physical geography of the region, to those rooted in an historical, cultural, or "the West as a state of mind" rhetoric. There is some agreement on what has been described as the "Unambiguous West" (Figure 1.1).[2] At the heart of the Unambiguous West are the Rocky Mountains. The eastern edge is defined by the semi-arid farmlands from Montana to New Mexico, while the Western edge is characterized by the Cascade and Sierra Nevada Mountain chains stretching from Washington to California.

The Unambiguous West is the "wild" cowboy West of the Madison Avenue Marlboro Man imagination. It is the interior American West, which historically has been perceived as a region dependent on resource extraction. This is dramatically shown in Figure 1.2. The Unambiguous West becomes "The Empty Quarter," or "The Marginal Interior." For Westerners, regional designations such as the "Empty Quarter" have a pejorative ring. This is not how they see the land of mountains, rivers, arid lands, and the wide-open spaces. But to outsiders this is the sparsely populated West, the image of the Old West that was built on extracting mineral, forest, and energy resources, and that was dependent on boom-and-bust cycles as recent as the 1970s and 1980s. Today the "Marginal Interi-

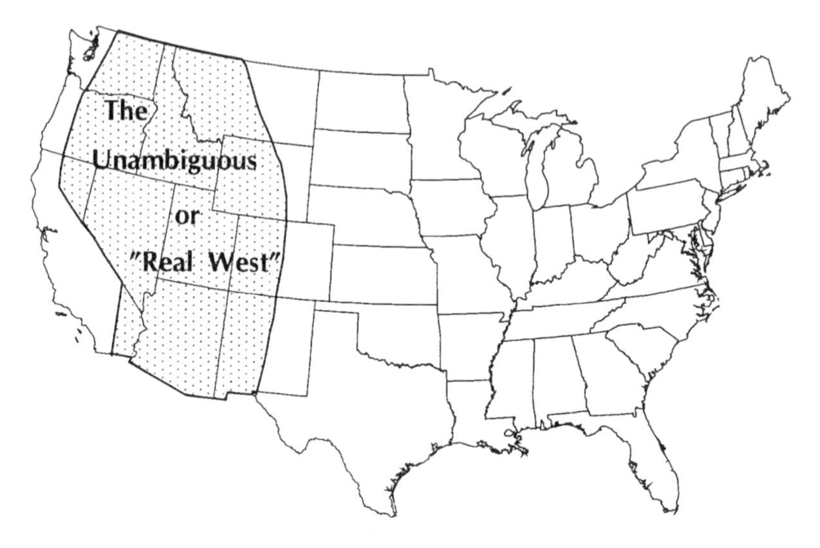


*Figure 1.1* The Real West? *Source:* Adapted from Nugent, 1992. See Note 2.

*Figure 1.2* The Marginal Interior. *Source:* Modified from H.J. DeBlij and Peter O. Muller, *Geography: Realms, Regions, and Concepts*, 7th ed., John Wiley & Sons, 1994, p. 210.

or" might better be called the "Amenity Interior," given the migration of people into the region seeking environmental and social relief.

The American West is defined by a climate of extremes, low population density, and boom-and-bust economies. In short, it can be defined by its physical geography. The Unambiguous West contains both the arid West and the well-watered Northwest. But can it be the West without Texas, land of cattle and oilmen? The extended West includes Texas, or West Texas at least. It extends to a Pacific West, which has at its core California, a state that evokes controversy about whether it is part of the "Real West." Still, for many people, what defines being in the West is the landscape, and access to it via wild public lands.

The myth of western opportunity also had a shaping impact on the region not only because it molded federal policy but also because newcomers brought their fears and mythic expectations with them to the West.[3] Today the West as a region remains to many the last great hope. The Empty Interior has become the "Last Frontier," "Last Escape," or the "Region of Hope and Optimism." It serves as a stark contrast to city streets where people are afraid to walk because of their perceptions of crime, increasingly violent gangs, drug dealing, and child prostitution. All of these become symbols of the failures of cities—especially large cities.

In the American West the large cities lie along the edges, either along the coast of the Pacific Ocean (for example, San Diego, Los Angeles, and Seattle), or at the edge of the Great Plains (Chicago), or along the Gulf Coast (Houston). With its vast interior and widely spaced cities, the people of the West have felt protected, but the diffusion or spreading of social problems often accompanies the new migrants as they seek America again in its western frontier.

Often what people are seeking is a life where they will feel more control over how they live and over their social, political, and physical environments, as well as an opportunity to create a place where they can be apart from the kind of society we have built in our cities. People moving to interior places in the West, particularly those moving from California, often have in common anti-urban feelings and a wish to detach themselves from endemic social problems.[4] Very few people today would turn to our cities as an example for other societies, given the perceptions, levels of fear, and stress attributed to living in such places.

Perhaps the rural West is one of the few places left to care about and love. Except for Woody Allen's view of New York, how many cities are portrayed lovingly? Large cities such as New York, Miami, Chicago, Detroit, and Los Angeles are seen as alien places where a concern for personal security prevails, and even post-modern portrayals of cities such as Los Angeles paint a relatively bleak picture.[5] The 1992 Los Angeles riots add to the impression and reality of hopelessness and despair for many people in large cities. This is in contrast to the optimism and high expectations of migrants to the rural West.[6]

Are people who are seeking a better quality of life migrating to one definable West, or to a number of different Wests? How many Wests are there? The definition of the American West itself as a region has changed over time. At first the West meant areas west of the Atlantic Ocean and closer to the Mississippi River; later it was any area west of the Mississippi; still later the 100th meridian was used as the demarcation point. As people moved westward, areas such as the Midwest gained their own regional identity. Many people exclude the westernmost part of California from the American West. Whether it is or is not a part of the "Real West," there is a circling back or eastward movement from the westernmost parts of California, Oregon, and Washington toward the inner West.

Regardless of where people draw the boundaries of the greater West, in actuality there are a number of different "Wests," each with their own set of characteristics. For example, the rainforest of the Olympic Peninsula of Washington, or northern Idaho and western Montana are clearly not part of the same region as southern Idaho, eastern Montana, or desert Utah with its canyon lands. In part this is a dry academic argument based on an old model of a region and preoccupied with the drawing of lines around an area or territory. There are different ways of defining regions.

A classic way the West was defined was as America's frontier. Frederick Jackson Turner imbedded in American history the frontier hypothesis. For Turner, the westward march across America resulted in pushing the frontier, the meeting place between savagery and civilization, forward until 1890, when he declared the frontier closed:

> And now, four centuries from the discovery of America, at the end of a hundred years of life under the Constitution, the frontier has gone, and with its going has closed the first period of American history.[7]

Turner closed the frontier based on U.S. Census estimates of population densities. The frontier was defined as having large areas with white population densities of less than two people per square mile. A map of the West in 1890 showed it dotted with areas having more than two people per square mile; both isolated and contiguous bodies of settlement were scattered throughout the West. So Turner and the U.S. Census Bureau declared the end of a frontier in America.

Counties with less than two people did not disappear even though the frontier did, as Dayton Duncan discovered when he went out to find such places and the people living in them. Duncan had no difficulty finding sparsely populated countries, as he reported in his fascinating book *Miles From Nowhere: In Search of the American Frontier*. Over a hundred years after the frontier closed, the 1990 U.S. Census listed 132 counties (outside of Alaska and in the Western states) with less than two people per square mile, with about one-half million people living in an area of over 400,000 square miles scattered throughout the West, and comprising about 13 percent of the lower 48 states.[8]

Many of the places that had less than two people per square mile in 1880 still had the same small populations in 1990. Most of the places are in the inner West, comprising most of the desert lands of eastern Oregon and Utah, almost all of Nevada, almost all of central Idaho, southwestern Montana, and a scattering throughout the rest of the Rocky Mountain states. There are no such counties in Washington and only a handful in California around Death Valley.

Many of these counties contain federal lands, and also federally designated wilderness areas. For example, the central Idaho counties contain the Frank Church River of No Return Wilderness, the largest such area in the continental United States. Even within the lower 48 states, there are still areas where people literally have not left their footprints. Nor will these places face the onslaught of urbanization, except around their edges, because of their remoteness and their designation as federal lands. They are slivers of wildness in a rapidly urbanizing West.

Another dominant theme of Western historians has been of the West as a region exploited for its resources by Eastern businessmen and bankers.[9] This enduring theme portrays the West as a colony of the rest of the country. More recently there have been fears that it would become a Zone of Sacrifice, a place to bury nuclear waste and other hazardous products, or dispose of other by-products of our industrial and technological society. The low densities—not just less than two people per mile, but less than five or ten per mile—mean that there are a large number of areas where there simply are not many people.

The extent to which exploitation, corporate control of resources, and other forms of economic control distinguishes the West historically from other regions is a debatable question.[10] Is it that much different from the history of control by mining companies in West Virginia and western Pennsylvania, or garment sweatshops in the East? Perhaps the perceived control by corporations headquartered elsewhere or by the "big government" in Washington, D.C. has lingered longer in the West or people are more aware of it. In other regions of the country, there is a clear history of labor-company disputes. The same has been true for the early organization of labor unions and the reaction of mining and other resource industries. The history is not a pretty one, with its accounts of violence against workers and federal troops being called in to put down state-declared "rebellions." This somewhat neglected history is similar to worker-union and corporate disputes elsewhere.

Where the American West differs from such current disputes as those involving autoworkers in Michigan or textile workers in the South is that labor is not fighting with corporations. In the West almost all of the hostility toward outside control is aimed at the federal government, which owns and manages the public lands, and not at the large corporations that work the land and over the years have employed fewer and fewer workers. As labor vents its spleen at the federal government, the corporations stand aside, agreeing with labor that the government is to blame. It is an old ploy, and one that plays well in the rural West.

The government plays a vital role in defining the geography of the West. It is these public lands that cannot be excluded from anybody's definition of the region. Approaching from the east, the West starts for me when the open spaces fill up with the sight and feel of mountains on the horizon. A sense of freedom, optimism, and Americana becomes overwhelming. Approaching from the west, from Portland or Seattle, it is clear that you are leaving an urbanized America as the landscape becomes dotted with small towns, and the larger ones take on a 1950s appearance. Everything slows down, and again space and distances prevail.

From whatever direction you approach the heart of the West you pass through the people's, or federal, lands. Some are marked by signs as you enter and leave them; others are not. But it is hard to drive through the West without feeling "this belongs to me." The towns and ranches that used to be considered the wildlands of the West are interspersed and dwarfed by the vastness of the public lands. Today they are tamed, categorized, and allowed degrees of wildness, with wildness meaning freedom from human intervention and consumption.

Wildlands define the West in a way that is different from the way that the largely private lands define the states from Maine to Florida, or the agricultural plains define the Midwestern states, or even its vast expanses define Texas. This raises the question of what is wild, and why did wilderness lands need to be designated in the first place? Are not the public lands of the West wild enough, whatever their designation? At one time perhaps this was the case, but at least since the end of World War II the agencies managing public lands have been driven by a need to produce commodities from these lands, whether timber, oil, minerals, cattle, sheep, or wildlife that was hunted for food or sport.

The land agencies were caught off guard as citizen pressures for noncommodity uses such as hiking, backpacking, preservation, habitat protection for spotted owls, salmon and other wildlife—endangered or not—created a demand for wilder public lands. Public lands had become production oriented and had become subject to the economic calculus of maximization of output. Wildness of public lands suffered and out of this emerged a movement to preserve some of these lands as wilderness.

## WILDERNESS AND AMERICAN MYTHS

Wilderness has served as the backbone of a mixture of American myths and reality. The early white pioneers perceived wilderness as frightening, as a land of "darkness" that had to be overcome, subdued, and conquered by the brave European settlers. In the path of these pioneers lay not only wildlands, but savages who also had to be conquered and then removed for the benefit of progress. The wildlands that the savages lived in had to be tamed by farmers, ranchers, foresters, miners, and others who would convert the wilderness to a better use of the land. Western American history became dominated by the exploits of the brave and, by most accounts, by men who both tamed and shaped its wild landscapes.

These accounts of the settling of the American West have become a part of the mythical and legendary America. And not just in America, but throughout the world. For example, Europe has many clubs whose members on weekends dress up in Western attire (or at least what they believe to be Western attire), play cowboys and Indians, and re-create the settling of the American frontier. Even in Japan, corporate executives and others will put on their cowboy hats and all the trimmings and go hang out in cowboy bars that look like they were taken from a set for "Gunsmoke" or a Clint Eastwood spaghetti Western. My students always show disbelief when I tell them about this, and they are bemused when they see on film Japanese "cowboys" talking about Western individuality and the freedom of the open range. Being a cowboy represents a small rebellion against conformity, group control, and doing what is expected, whether in Japan or elsewhere.

The wildness of the West is important, and not just to nostalgic Americans. The West and its cowboys have become legendary throughout the world, partly because of translations of the writings of authors such as James Fenimore Cooper and Louis L'Amour. Europe and other places in the world have spawned their own writers who have written and continue to write about its unique past. The American West has often been portrayed by writers outside the United States as a sort of utopia. The mythical West was the place where wilderness and civilization clashed. It was a world where male heroes of the highest moral character and manliness clashed with the wilderness and savages. Out of that came the new man who helped to carve and develop a better and more developed democratic society. That was how it was according to America's leading historian Frederick Jackson Turner. Both American and European writers romanticized this binding up of wilderness as forming the character and destiny of the new nation.[11]

Europe was seen as the old decadent civilization, America as the country of the future. The best known of European writers on the American West was the German Karl May, whose books sold more than 40 million copies. His books revolved around a German writer in the West and a noble Apache Indian, Winnetou. His stories were actually rooted in German national myths transposed to the American West, with the traditional themes of good versus evil and the spreading of Christianity. His novels were not realistic in detail because he never visited America until four years before his death and after selling millions of his books, many of which were made into films. Most Americans have probably never heard his name, but as a German friend told me, "I spent my childhood growing up on Karl May books, and dreaming of the West."[12]

The image of the West even remained alive and strong behind the Iron Curtain of communism. A graduate student of mine grew up in Estonia in the former Soviet Union. Estonia, which was an independent state between World Wars I and II, was a part of the Soviet Union until 1991 when it regained independence. We decided to look at the perception of Estonian university students of the American West—where it was, how well they could draw it on a map, and what characteristics they ascribed to it.[13] I found the results very striking. Not only could the surveyed students draw accurate maps of the American West, their images of it were based on the physical geography of the area. These were students who obviously did not get much emphasis on this region in their Soviet studies. Schooling in the former Soviet Union stressed the negative aspects of life in the United States, and certainly did not portray the American West as a mythical utopia of individualism and freedom. The Estonian students, given the similarity of their language to Finnish and the proximity of Estonia to Finland, did have the advantage of being able to receive and understand their television broadcasts. What struck me as fascinating was their definition of the West by the natural surroundings. Despite the prevalence of images from Finnish television of the modern Urban West being of Houston or Los Angeles, for them the West was defined primarily by a wild landscape.

Whether in free Europe or behind the Iron Curtain, the mythology and image of the American West has remained powerful. Unfortunately, the Mythical West was not the Real West. Even the portrayal of this Real West came under attack, most notably by the "new historians." The "New Western History" has charged that the Utopian West was not only a myth, but that Western history, as described by Frederick Jackson Turner and subsequent historians, was mostly an ethnocentric and biased interpretation of the settling of the West.

## THE REAL AMERICAN WEST, PAST AND PRESENT

The attack against Turner's version of western history was lead by Patricia Nelson Limerick, whose 1987 book, *The Legacy of Conquest: The Unbroken Past of the American West,* probably dealt what should have been a fatal blow to his frontier hypothesis. In Limerick's view, Turner was ethnocentric, racist, and nationalistic. English-speaking white men were the stars of the Western pageant, a tradition which has too often been continued.[14] For example, the Texas Rangers acquired an aura of myth; they were idolized and exalted as true heroic figures in American history, and popularized in modern times in a television series and movies. This popular image is not enthusiastically embraced by Mexican-Americans who shared Western spaces with these noble white men.

Historian Susan Armitage says that Limerick's book was very courageous, because until 1987 most Western historians were unable to fully depart from our most famous image of the frontier experience as "fundamental to the American character," and our most cherished myth, which most Americans believe in even today: "that it was the frontier that made us the freedom loving, democratic and optimistic people that we are."[15]

The new Western history de-emphasized the frontier, and its supposed end. Instead, the West became a meeting ground for Indian America, Latin America, Anglo-America, Afro-America, and Asia-America. In many places Western people were groups of strangers and the frontier came to no smooth end. Today, all these diverse groups share the same region and its history, but they wait to be introduced.[16]

The "old" Western history had a historically diverse cast of heroic characters, but they had in common one distinguishing characteristic—they were all men. Sue Armitage has called the American West "Hisland" because the role of women often has been trivialized and ignored. Critics of past interpretations of the West argue that the importance of including women is not simply to add more historical actors, but to gain a new perspective, remove the distortions of the past, and ultimately present a more accurate and probably a very different view of the past.[17]

If the role of women and of racial and ethnic groups has been ignored, even more evident has been the denial of the Native American influence on our cultural landscape. Turner either ignored Native Americans, or viewed them through the racist prism of his time as savages unable to contribute toward a civilized society. Therefore, it is doubly ironic that today the Native American view of a matriarchal society and a reverence for Mother Earth is a model that is often held up as an alternative example of how we could become a more sustainable and just society. There is also increasing recognition that Native Americans have already had a significant impact upon the dominant culture of the United States.

The desire to preserve wilderness and stop destructive practices on our public lands is one indication of Native American influence. The search for amenities and a clean environment in the rural West by the "new migrants" reflects in part the search for places connected to, yet apart from, the results of an exploitive, efficient society that isolates and segregates those who do not fit in. The landscape of the past, of the American Indian, whether romanticized or not, better fits the image of what the new, mostly white migrants seek in the West.[18]

Some of the reaction against the new interpretations of the American West indicates a fear that the EuroAmericans will become cast as the "bad guys" instead of the heroes they were formerly portrayed to be, and that would undermine the myths they created for the rest of us.[19] But is ignoring reality or whitewashing the past better than making the stories of Western history more interesting and inclusive? This is particularly important when we consider the role of wilderness and our public lands as we struggle to "manage" them better. The wild landscape has played a critical role in shaping society in the American West. A cultural landscape rooted in a frontier mythology that extolls freedom but is drenched in the blood of conquest has slowly begun to change toward a recognition of wilderness as a place worthy of being left alone, and of being preserved for its own sake as well as ours.

In many ways, we Americans deceive ourselves when we see ourselves as the most advanced nation on earth, for we are still an immature culture, a society all too close to its pioneer stage of leveling forests, mining resources, and amassing wealth at the expense of our ecological and social fabric. That is why we need to consider the accuracy of our history. Landscapes are etched by historical processes. The old forms are not discarded, but retained. In the West the dreams and myths remain, but their accuracy comes under scrutiny. Public policy decisions, particularly of how our wildlands are managed, that are based on historical inaccuracy may continue to legitimize actions rooted in inaccuracy, sexism, or racism. For example, many of the conflicts of the past are coming back to haunt the conquerors as the rights to land and resources are "returned" to Native Americans and Mexican Americans by the courts. Ironically, the conquerors see themselves as victims, prevented from access to or development of resources as they deem fit.

The geography of our public wildlands, water, and Indian reservations are among the features that both define the West and are an intricate part of developmental conflicts. Many of these conflicts are an outgrowth of federal policies (mainly subsidies). Americans did not realize that setting aside public lands, water projects, crop subsidies, and passing a variety of laws that fostered the extraction of resources from wildlands would take on a life of their own and create new landscapes. Migration, in particular, established an entirely new geography of the West.

Many of the natural resource laws operating today are a product of the need to settle the frontier and civilize both nature and the people living there. Today, as exemplified by the new Western historians and others, our view of the past is changing, as are our attitudes of our relationship to nature and toward our public wildlands. The laws embodying the old beliefs remain, old outmoded laws that have led to pitched battles between environmentally oriented migrants, residents, and Westerners who are trying to preserve economic interests inherited from laws and mores of the past.[20] For example, wilderness advocates and loggers contend for the high ground of what our Western landscape should look like. Too often the multiple layers of a Western history come peeking through in ways too reminiscent of a past that is only now being better understood and openly discussed. Recent threats of violence against land managers and the bombing of their offices casts a shadow both on our present and past.

## ROOTS OF THE AMERICAN WILDERNESS TRADITION

The way people view wilderness emerged out of basically two different traditions. One came from the heads of writers and artists and was often tinged with romanticism. This tradition views wilderness as essential to all people. It is the home from which we all emerged, and with which we still retain primordial ties. Even in a highly technological world, we will suffer if our bonds with nature are severed completely.[21]

This attachment to nature and the landscape we inhabit need not be limited to rural wild places. It is through this attachment to landscapes, whether in Brooklyn or in Wyoming, that we develop a sense of place and of belonging that is vital to our well-being. Therefore, the more we diminish the wild world around us, the more harm we do to ourselves not only individually, but also collectively as a society.[22] Thoreau's famous words, "in wildness is the salvation of the world," continue to ring true when we hear them. From Thoreau and others who followed, most notably John Muir and David Brower, the message was that to diminish the wild creates an emptiness that modern society cannot fill.

The other strand took a more scientific and rational approach, focusing on the interrelatedness of nature and the view that to greatly disturb or modify nature had negative consequences for society as a whole. It has built on the observations of George Perkins Marsh, who, in the 1850s, wrote about the possible negative consequences of the continuing industrialization of the United States. His observations were based on his experience as U.S. Ambassador to

Italy. He described the ecological destruction of Europe, which occurred largely as the result of deforestation, and he warned of the severe consequences of the rapacious deforestation taking place in the United States. His book was a best seller and a precursor about the ecological impact of industrialization long before the word "ecology" had been coined.[23] In the United States he was followed by others we shall hear more from, such as Gifford Pinchot, the patron saint of the U.S. Forest Service, and Aldo Leopold, who in the 1930s promoted the setting aside of wilderness areas from his position within the Forest Service.

Both within and outside the American West, wilderness is increasingly valued for aesthetic-spiritual values as well as for providing a myth-laden cultural identity for the West. There has been a dramatic shift toward viewing wilderness not as a fearful landscape, but as one that needs to be protected because it and the endangered species in it are in danger of disappearing. Indeed, efforts are underway to make the landscape of the West more "wild." The idea that there is not enough "wild" in the American West is quite a change from the way both the Mythical and Real West of our past were viewed. The type of Western landscape that emerges in the future will define whether the West as a region will remain more or less unique. Part of this future has been entrusted to land managers in agencies which to most Americans have been obscure and out of the public eye.

## MANAGING THE PEOPLE'S WILDLANDS

America's public wildlands have become a managed landscape. Professionals in various federal agencies have tried to control and manage what is to remain wild and what is to be developed for commercial uses. For some, wilderness is an inauthentic landscape or, as some bumper stickers boldly proclaim, "Wilderness—Land of No Use." The landscape of the West requires that workers be allowed to use federal lands to extract a living, by felling trees, mining minerals, and grazing cattle or sheep. People who use public resources in their work continue to assure everyone that they are concerned about the best uses of the land both now and for the future.

The debate over managing the land often is framed in simple terms, such as a contest between Yuppies and "real people," or a War for the West between cappuccino and light beer drinkers. To some the choices are clear, an Old West or a New West, each of which represents different views and attitudes. Take your pick. You can have one or the other. Each choice infers a different set of policies toward land management, and the agencies are caught in the middle of a crossfire. In reality the choices are not so clear-cut, and the changing West is a much more complex place than either of the proposed choices.

Unfortunately, the debate has been a boisterous one, filled with truths, half-truths, and outright lies. As a nation we need to get beyond this. To do so we must understand how and why our public lands define the West as a region and a way of life. We must understand current changes by referring to the context of the past. Change has been characteristic of the West. Just how the American West continues to change and why is part of the ongoing debate.

A basic theme is the change toward viewing wilderness as an amenity landscape, and the consequences of that change for how federal lands are managed. There has been a dramatic shift in the role of public lands, from allowing their use by people who have been raised with a modern-day frontier and cowboy culture mentality to providing for the demands of urban and suburban users. The movement of these people into and around federal lands heightens the cry for nonconsumptive uses of these lands. This has a profound impact on what the West will be like in the future.

Efforts to stem the tide of change will be fruitless, though conflicts will continue to fix national attention on these changes. The debates over wilderness and other lands may appear to be new as people strive to hold on to the old ways, and corporate and political interests spend money and exert influence to prevent change, but they are not. That is why the past must be reckoned with as it keeps pushing itself into present debates, whether in the form of the Wise Use movement or—in a more sinister fashion—hate groups, militias, and others trying to find some defensible space.

The public land managers are caught in the middle of these changes. The control and management of these lands are slowly being wrested away. Increasingly, decisions are being determined in the courts. Trust in land managers has plummeted. How this happened is a story that is just starting to be told.

Conflicts over the use of wilderness have been especially contentious in places where the land management agencies have subsidized a way of life and participated in local boom-and-bust cycles. Areas where people have built a way of life around harvesting trees or running cattle on public lands are under siege. The managing agencies bear responsibility because they have followed policies in the past that are unsustainable and destructive of the environment.

Managing wilderness is not easily compatible with providing extractive or agriculturally based jobs. Nor is this the image agencies such as the Forest Service have tried to portray with their Smokey the Bear campaigns that have reached into urban homes via television. Smokey presented the image of an agency working to protect forests and wildlife from fire, thereby preserving habitat. Whether intended or not, this image was misleading, since during that period the shift was toward extraction of forest products to supply private industry and an increasing demand for housing, not toward habitat protection.

As an agricultural agency, the Forest Service has proved to be quite efficient, but as an agency protecting habitat and wildlife it has been lacking. The record of the other federal land management agencies also is spotty on protection of wildlife and habitat. While it is easy to blame the agencies, this is largely a form of scapegoating or Monday morning quarterbacking. Though headquartered in Washington, D.C., the agencies are rooted in the West, and have consistently been responsive to the demands of local people and Western political representatives in Congress.

Though used mostly by people living near them, public lands and especially wilderness belong to the general public that subsidizes the many benefits local residents get from them. Historically, local and state residents have had a greater say in how public lands should be used. The designation of wilderness was a recognition that such management strategies were not reflective of a larger public interest. This struggle between local users and national interests is continuing. At times and in specific places it has been muted by the continuing migration of people into the "Real, Unambiguous, or New" West. In other places, the lines of battle have been drawn, and the use of descriptions such as "the War on the West" make obvious how some feel about the changes taking place. Wilderness both directly and indirectly is at the heart of this conflict.

The management agencies find themselves swinging back and forth between the different demands of these groups or trying to strike a down-the-middle compromise. This has not worked well. What is needed is an understanding of how to move diverse groups with an affinity for the land and wilderness beyond the stereotypes they hold of each other. The continued inability of the managing agencies to do so has become more and more obvious. These problems have a history. Unfortunately, the debate over setting aside wilderness areas and how they should be managed set in motion many of the existing conflicts.

*Chapter Two*

# History and Management of Wilderness

There has been little consensus and much debate about how "wild" wilderness and other public lands should be. Should they be wild enough to have wolves and grizzlies? Should they be restored to some pristine condition? Should all evidences of man's past intrusion be obliterated? How much intrusion by brightly colored backpackers outfitted with gear and clothing from upscale merchandisers should be allowed? How many horses and their droppings should there be along the trail? Should mountain lakes be stocked with fish? There are any number of questions whose answers impact on how "wild" our protected wilderness will be.

There is no clear philosophy to guide the management of wilderness despite the 1964 Wilderness Act. There are four different federal agencies charged with designating, protecting, and managing these lands. These agencies all manage wilderness in different ways. How they manage these wildlands influences the lives of people and the economies of localities in the American West.

Among the questions these agencies have handled differently are how much of the federal lands should be allocated as wilderness, and how should questions about allocation be decided?

Because the agencies control the review process, and are responsive to local and Congressional political pressures, the "scientific" basis of making decisions has become suspect. The agencies have come increasingly under attack by various organizations claiming to represent the "public interest." To understand how and why this has happened, a short history of these agencies is in order.

## A SHORT HISTORY OF WILDERNESS DESIGNATION

The 1964 Wilderness Act established 9 million acres of wilderness lands. Today there are about 100 million acres of wilderness. This sounds substantial. Yet it is less than 3 percent of the total United States lands. Considering only federal lands, wilderness accounts for about 15 percent of all our public lands. But this 15 percent seems to generate an inordinate amount of controversy, given that over 85 percent of federal lands are not designated as "wild."

The cause of the controversy surrounding wilderness areas is not that they contain a lot of resources. They were, in part, designated wilderness because they were not considered to have much current commercial value in terms of timber, minerals, or oil and gas. If heavily forested, the costs of cutting the timber in these roadless areas are often quite high and not profitable for private companies. Also many of these areas have not been explored for minerals, making it difficult to know if they do contain resources. Even if they do, wilderness designation and resource extraction are not considered compatible activities. Many wilderness areas are largely rock and ice, and though they are quite scenic, they represent only a fraction of the type of ecosystems found on federal lands.

When Europeans settled the United States, wilderness was considered an obstacle to be conquered, with little or no value. Land was abundant, and the "frontier ethic" prevailed. With growth and development, a continent once entirely wild was almost completely subjugated. One consequence was the logging over of much of the original forestlands.[1] Wilderness was no longer in super-abundance.

The increasing scarcity of wilderness led a number of individuals and organizations to try to preserve such areas. As early as 1921, there was a proposal for a nationwide system of wilderness areas, but it was not until 1964 that a Wilderness Act was passed. The campaign for wilderness was pushed by preservationists and recreationists, who did not trust the management practices of the federal land management agencies, and was opposed by grazing, timber, mining, and irrigation commodity interest groups. Initially, both the U.S. Forest Service (USFS) and the National Park Service (NPS) opposed the designation of wilderness areas. The road to final passage of a bill was contentious, and it took seven years to achieve final passage of the legislation in 1964.[2]

The 1964 Wilderness Act established 54 units covering just over 9 million acres. Since then the amount of land in wilderness has increased tenfold to about 100 million acres. This figure is a little misleading since one state, Alaska, has the majority of wilderness lands with over 60 percent of the total wilderness acreage. The remainder is distributed over 43 states, with the Western states having the predominant share (Figure 2.1).

The debate over how much wilderness there should be will continue and, given current trends, ultimately somewhere between 100 and 125 million acres will be federally designated wilderness. Most of the present wilderness acreage was set aside after public hearings and recommendations to Congress from state delegations and various interest groups. In some states this process has not worked, and conflicts over how much land should be designated wilderness continues unabated. If political settlements cannot be worked out in Congress, the debate shifts to each of the national forests as individual plans are challenged by interest groups. The Forest Service has both the most units and acreage in wilderness in the continental United States, but the National Park Service, the Bureau of Land Management, and the Fish and Wildlife Service also manage wilderness. The management often is of a patchwork of adjacent or intermingled wilderness lands administered by these four agencies, or even by different land managers within the same agency.[3]

## MANAGING WILDERNESS

Managers of public lands have come increasingly under attack. To understand why, it is important to realize that most of the public lands are not managed to preserve wilderness, but to provide com-

***Figure 2.1*** Federal Wilderness Areas. *Source:* Adapted from Charles I. Zinser, *Outdoor Recreation: United States National Parks, Forests, and Public Lands* (New York: John Wiley & Sons, 1995).

modity resources. The national forestlands that contain most of the wilderness in the continental United States originally were set aside because of concern about the timber shortages brought on by the private pillage of woodlands that accompanied the nation's westward expansion.

Until relatively recently, land managers of the national forests were trying to accommodate timber production, grazing interests, and watershed protection; preservation of wildland ecosystems was not considered a priority. Increasingly there has been growing opposition to the extraction of resources from the public lands. The timber-harvesting policies of the Forest Service have been challenged by preservationists and the general public. This has posed special problems for land managers, who generally see timber harvesting as a more important use of the forest than noncommodity uses, which they feel place undue constraints on harvest practices. Many regard the designation of wilderness areas, where even recreation is limited, as absurd.[4] Wilderness management and associated funding has a low priority within the public agencies. The consequences have been summarized by prominent conservation writer Michael Frome:

> I daresay that I've been to as many wilderness areas in this country as anyone, but I haven't seen a single wilderness anywhere in this country managed as it should be. . . . I've seen wilderness areas in terrible condition, abused and degraded, often as not by uncontrolled and inappropriate recreation use, getting worse rather than better, staffed by inadequate personnel insufficiently trained. If you ask me, the shabby state of our wilderness today is not a reflection of good forestry, nor good park practice, nor good wildlife management, nor good rangeland administration.[5]

Some of the antipathy toward wilderness by federal managers can be explained by the history of the managing agencies. The National Park Service considered their parks to be wilderness that was managed adequately and without need of being given any new designations. The increasing commercialization of the national parks, with its emphasis on attracting and catering to an automobile-based visitor and the resultant overcrowding, seems to belie them as places that emphasize wilderness.

Agencies are made up of people. Within the USFS and other agencies there were only a handful of people early on who argued for the setting aside of wilderness lands. Aldo Leopold is perhaps the best known of these early wilderness advocates because of his conception of establishing primitive areas in the national forests that would not be managed for commodity production of any kind. He and a few others were the exception to the rule.

For the most part, federal land managers saw it as their role to follow good "scientific" principles as they managed the lands. The manager was to be practical, rational, and ever efficient in managing the lands for the public good. The problem was, and remains, what is the public good?

The agencies, and especially the USFS, answered this question by referring to the utilitarian philosophy of "the greatest good for the greatest number." The USFS decided that providing for a sustained yield of trees while minimizing harm to the forests was the best way to serve the "public interest." This was the approach promoted by Gifford Pinchot, who was a founder of the Forest Service and whose spirit continues to hover over the USFS today. Pinchot's name is often invoked when it comes to justifying the practices of today's Forest Service. Whether this is fair can be disputed, because Pinchot was a liberal creature of his political times, and where he would stand today could be endlessly and needlessly debated.[6]

Pinchot was trained as a forester in Germany and designed the USFS after the German model of forestry. The German model in many aspects was based on the Prussian army and therefore is organized in a top down and hierarchical fashion. In the USFS, authority is delegated from a centralized management authority in Washington, D.C. As in the military, the field foresters are expected to respond as good soldiers would to the dictates from above. They develop an allegiance to the organization, but at the same time are expected to move to various assignments throughout the United States. They view themselves as foresters whose duty it is to carry out the mission to manage and protect the public lands under their care. Like any soldiers working to protect the public, they expect public support. In turn, the public looks to and depends on them to do the "right thing" and does not expect to get involved with how they go about carrying out their responsibilities. This worked well during Pinchot's reign. The USFS was the protector of the forests and its wildlife from the forces of the marketplace that would lead to their destruction.

Pinchot wanted to wrest timber production from private corporate interests that otherwise would clearcut their way across the nation in an unsustainable fashion, leading to timber shortages. The forests of the East and Midwest were rapidly disappearing and there was fear that a similar denuded landscape would characterize the American West. With Pinchot's urging, President Theodore Roosevelt expanded greatly the national forests. Literally under the cover of night, they pored over maps and, using emergency powers granted to the president, set aside vast acreages in the West that later would become the core of the USFS. Under Pinchot's guidance the USFS would manage the people's forests in a sustainable way, yet still provide the needed supply of timber. The management of the U.S. federal forests would show the world how forests and landscapes should be managed.

Ironically, since the 1970s, while swearing to uphold the traditions established by Pinchot, the USFS has increasingly been accused of promoting and using clearcutting and other unsustainable practices to overcut the American West. The USFS, formed as a reaction to the damage done to the American landscape by corporate practices, has had to defend itself in the eyes of the public for using corporate logging practices. It has even been accused of being an unofficial arm of corporate timber companies. The battles have usually been centered on the Western landscape.

One view was that the result of national forest management was a fragmented landscape littered with more roads than the interstate highway system. Clearcutting and other practices had created a cutover landscape with silted streams, poisoned trees, and endangered wildlife—an American Amazon in the making. Environmental books showcased clearcutting with aerial and satellite photographs displaying a doughnut-hole riddled landscape. Some charged that the USFS could not have done a worse job if they had tried.

The USFS responded as though it had been stabbed in the back, misunderstood, and spurned unfairly. The attacks were overblown, misleading, and unfair. A forester was an environmentalist for whom every day was Earth Day. Not all clearcutting was bad; wildlife actually benefited from the harvesting policies. Silted streams and disappearing salmon were not related to forestry policies. USFS employees began referring to themselves as "combat foresters" or "combat biologists." The military allusions are unmistakable. The loyal troops are under attack, first by "extremists" and then increasingly by a public that does not understand them.[7]

The transition from a populist agency striving to show how it could protect and manage the people's land to one increasingly identified as a government-owned forest corporation can be traced partly to an increasing emphasis on efficiency and to Pinchot's attitude that the manager knows best.[8] The manager is a rational and efficient overseer and public servant working for the public interest. The catch is that the manager knows and determines what is in the public interest and then tells the public.

This is a dilemma in all of the land-managing agencies. They manage for the public good, which is defined vaguely at best or not at all. A charismatic and politically influential leader such as Pinchot sets the direction and defines the public will.[9] There has been no comparable charismatic figure leading the Forest Service since Pinchot. Historian Paul Hirt, in a critical look at the agency, argues that it tried to be all things to all people and got caught up in a conspiracy of optimism. Ultimately, its optimism failed, and, when confronted with a dual mandate of production and preservation, the agency opted for both with more intensive management. The "conspiracy of optimism failed as these high levels of resource extraction began destroying forest ecosystems."[10]

The history of the land agencies contains little scrutinizing from the public eye (barring an occasional political scandal), and shows that they have been able to pursue their objectives as they saw fit. The NPS and the USFS developed public relations campaigns to educate the public about what lands they should use and how they should use them: visit the parks and keep the forests safe from fires. These slogans and admonitions affected us during childhood, setting the tone for what we considered wild. Many of us were influenced by playful or thoughtful cartoonized and humanized bears such as Yogi and Smokey the Bear.

Today we seem to be caught in a time warp. Yogi still creates lovable problems for the park rangers in cartoon reruns. Smokey the Bear is in mid-life crisis and semi-retirement not sure that putting out all fires is a good idea. Even in the 1960s experts were disputing the wisdom of the USFS policy of stamping out all forest fires.[11] The USFS convinced millions of us that putting out fires was one of the heroic and noble duties of forest rangers. No longer.

The extinction of all fires has been shown to be a flawed policy. Instead, wildfires are allowed to burn unhindered in wilderness areas. Many other forests are considered to be in danger of having large fires from the buildup of flammable material on forest floors from years of suppressing all fires, natural and human-induced, deliberate or accidental.

The USFS is promoting salvage logging to "clear out" the dangerous wood that has built up from years of putting out fires. They are facing fierce opposition in some areas. The public is no longer passive; it is unwilling to take statements at face value from land managers who say that they know best. Debates rage over whether salvage logging is being promoted to provide trees for the wood products industry or to improve the ecological integrity of the forests. Skeptics abound. Landscape ecologists argue that dead and dying trees should not be harvested and are necessary for the ecological integrity of the forests.

Land managers feel besieged and try where possible to reach some kind of compromise. Where compromise is politically unpalatable, land managers may be forced to accept decisions from superiors up the chain of command who too often are responding to Congressional pressures. Western Senators and Congressmen historically have put pressures on land-managing agencies to provide commodities from the public lands. The Western public lands stand as a warehouse from which industry can withdraw trees, minerals, energy, and other commodities at bargain prices.[12] These commodities are marked up, processed, and sold back to the very public that owns the warehouse. As a result, communities founded and dependent on such withdrawals have sprung up throughout the West. People dependent on jobs or profits from commodity extraction put pressure on their elected representatives to keep the supply forthcoming.

Environmental organizations as well as individuals respond with demands for a more open process and often sue for enforcement of existing laws and regulations. They want a halt to what they see as giving away at low prices not just commodities, but portions of ecosystems piece by piece. The continuing confrontations between land managers and citizen and environmental groups as evidenced, for example, by legal challenges to almost every forest plan proposed by the Forest Service, has forced a reconsideration of the criteria under which public land managers operate. Federal land managers may simply be trying to satisfy too many groups with too general a Congressional mandate.

## MANAGING WILDERNESS WITH THE MULTIPLE USE CONCEPT

Multiple use is a guiding principle of resource managers. Unfortunately, multiple use is not an appropriate tool. Agencies set up to protect and manage lands for multiple use actually had a pro-development bias. With few exceptions, that philosophy has remained dominant to the present day.[13]

When Gifford Pinchot introduced the concept of multiple use, he was trying to accommodate production of timber, grazing interests, and watershed protection. Commodity production, and particularly tree harvesting, was to be the major function of the USFS. It was to be done in a scientific fashion to promote sustainable yields over a long period of time. Pinchot wanted to demonstrate that the timber mining done in the pursuit of corporate profits was not a proper way of managing forests. The federal forests were set aside to show that the government could and would manage forests better. Gifford Pinchot promoted federal management of the forests under the utilitarian mandate of "the greatest good for the greatest number." More trees for home building and other wood products over the longest time would benefit the public directly. If left solely to corporate management, lands would be cut over and there would be periodic and long-term timber shortages in the United States. By contrast, well-trained government foresters would manage harvests while protecting watersheds and the long-term viability of the forests.

Pinchot installed in the USFS an *esprit de corps* that continues today. Correct and sound tree harvesting will provide what Americans want and should have. Under the Pinchot doctrine, multiple use promotes timber harvesting within the constraining need to provide a long-term managed—but cut—forest. How and where patches of forests get cut at particular time intervals is a scientific decision.

Since Pinchot's era, the number of groups demanding equal consideration under the multiple use concept has increased greatly. The demands for noncommodity uses of the public lands have increased; setting aside wilderness is one result of such increased demands. Unfortunately, many managers are not well trained to deal with noncommodity uses and cling to their commodity bias. U.S. Forest Service managers sincerely believe that promoting timber harvesting is a more important use of the forest than recreation or wilderness preservation. In the opinion of many, to put other noncommodity uses above timber harvesting is to have the tail wagging the dog.[14]

Traditions by definition are inertia bound. Shifting toward managing an environmental rather than a commodity forest requires a change in priorities. Until recently forest management was less complicated because it was not in the public eye, nor was it an issue meriting national attention, to say nothing of causing controversy.

Until recently, little attention has been paid to how public lands were managed. Faith was placed in public land managers and their symbols, such as Smokey the Bear, to do the job right. Today, regional issues, whether controversies about owls, salmon, grizzly bears, or wolves, are really about how public lands should be managed and used. National attention has been focused on management decisions normally entrusted to agencies that, in the past, responded mainly to local and regional interests.

The nationalization of wilderness and other public land issues has put the focus on how that landscape is being managed. That scrutiny and public attention has shown the failings of the multiple use concept. Multiple use has been a guise used to allow public land managers to practice dominant use while professing to practice "progressive" management of Western lands. Unfortunately, the land-managing agencies increasingly have been identified as promoting corporate rather than public interest.

Wilderness and other public lands should be managed in the public interest rather than in the interest of private individuals, corporate entities, or the managing agencies. Multiple use does not accomplish this.

The concept of what is in the public interest has changed from Pinchot's era. Defining the public interest is not easy. The use of the public interest rather than the multiple use concept is more in line with Aldo Leopold's concept of a land ethic where natural resources exist and are important in their own right.[15] The land ethic is contained within a larger environmental ethic that recognizes the potential harm as well as benefits of a technological society.

The agencies managing wilderness areas consider them useful primarily for recreation, which is in line with their commodity-based approach to management. A wilderness area is useful if it attracts visitors. If the number of people visiting and "using" wilderness is small or declining, this is evidence that there is either enough or too much wilderness. Body counts become the most important criteria for measuring the usefulness of wilderness. Such management strategies are wrong. Attitudes toward wilderness within the agencies need to be changed, and the management structures as well.

## GENERAL MANAGEMENT: TOO MANY CHIEFS

A critical management problem is the patchwork quilt of adjacent or intermingled wilderness lands administered by different agencies, or even by different land managers within the same agency. Federal agencies that manage wilderness do not have similar philosophies and management strategies.[16]

The approximately 100 million acres of wilderness is managed by four different agencies. The largest amount of wilderness in the continental United States is managed by the U.S. Forest Service, which is within the Department of Agriculture. The other three agencies (Bureau of Land Management, National Park Service, and U.S. Fish and Wildlife Service) are in the Interior Department. They have not applied the same criteria to human use and intrusion into the wilderness. Managers in each of the agencies do not receive similar training and are not rewarded equitably. Similar management practices are not used to protect wilderness areas from external and internal threats. Interagency politics has become the order of the day.

How wilderness is managed can vary greatly even within the same agency. An internal task force concluded that management of individual wilderness areas of the National Park Service is not carried out on a systematic basis nationwide. They found this lack of consistency to be true for designated, potential, proposed, and defacto wilderness areas.

Laws and regulations already exist that can be used to protect wilderness. For example, under the Clean Air Amendments, the original wilderness areas were classified along with national parks as Class I areas. This designation essentially prevents any significant additional air pollution of these areas. The federal land manager also is required to protect values related to air quality such as visibility, and to prevent construction of a facility if such values are violated.

Given their training, federal land managers have not been eager to assume responsibility for enforcing pollution regulations as well as halting development outside federal wilderness areas, even if they have the power to do so.

## INTERAGENCY COOPERATION

One argument for having different agencies manage wilderness is that this creates competition and promotes efficiency. Each agency will try to outdo the other and promote innovative ways for protecting and managing wilderness. Although theoretically plausible, there is little evidence to back this up. Indeed, all of the agencies have been unenthusiastic about managing wilderness, often having to be forced by lawsuits and the courts to carry out their mandates.

Interagency cooperation often is promoted as a means of managing wilderness within a similar region. There has been a lot of rhetoric about interagency cooperation, but little substance to back it up. One of the best known of America's parks, Yellowstone, serves as a good example. Throughout the world, Yellowstone is considered to be one of America's national treasures. There has been a lot of concern about the need to consider the larger Yellowstone ecosystem if America's first park is to be preserved for future generations. It has been recognized that the Yellowstone ecosystem should be the basic unit for management and policy making.

The Greater Yellowstone ecosystem is managed by all four agencies mentioned above. Wilderness in the area is not confined to one agency. Even within one agency, the Forest Service, management authority is diffuse. The seven forests report to three separate regions headquartered in three different cities.[17]

If there is to be interagency cooperation, surely it will take place in Yellowstone. Not so. The Forest Service has a history of clearcutting up to the park boundaries, much to the distress of park managers, so much so that when it snows a clear boundary can be seen between where the Park Service lands end and the Forest Service lands begin by the absence of trees across the boundary. Further potential problems loom from the oil leases issued by the Forest Service on surrounding lands and ever present geothermal development threats.[18]

A cooperative approach would require working together to reconnect the Yellowstone ecosystem and to restore the landscape. This has not happened. Any cooperation that has taken place has occurred within a highly charged and politicized environment. Similar, less visible examples can be cited throughout the West. The North Cascades National Park east of Seattle is surrounded by national forests with a similar history of resource development outside the park. I recall the irony of hiking out of the wilderness in the Gifford Pinchot National Forest and being struck with a sense of being surrounded by clearcuts and dodging logging trucks. Is this the appropriate legacy for Pinchot?

If agencies do not work together and cooperate well in such a highly visible environment as Yellowstone or the Cascades, can they be expected to do so in the more isolated Western areas containing the bulk of wilderness lands? The history of the interaction between the Department of Agriculture and the Interior Department is not replete with examples of cooperation. Having these different departments and agencies managing overlapping and complementary wilderness areas and ecosystems only makes the situation more unstable.

As illustrated by the Yellowstone example, the Forest Service in the Department of Agriculture has cooperated with other agencies in the Department of Interior, such as the Park Service, at best only in manners of public relations and not substance.

This division of land management is the result of historical events, not a logical way of managing public lands. It goes back to Gifford Pinchot and his desire for independence in managing national forests. What may have been rooted in the political realities of the nineteenth century is no longer rational and in the public interest today. However, all is not doom and gloom. There have been, and continue to be, isolated examples of different agencies cooperating and working together. Nonetheless, it is naive to expect that even in such examples the arrangements made will necessarily continue indefinitely into the future.

A history of inconsistent management exists even within single agencies. In all the Western states there is more than one national forest. Often they are adjacent to each other. Yet depending upon their management philosophies and strategies, managers of the different forests may view the role of commodity extraction, biodiversity, recreation, and wilderness quite differently.

The local politics may differ as well. Communities historically built around logging, mining, or ranching will be steeped more in tradition and myths than either more diversified communities or those attracting recreationalists. They and their local elected officials will be more rooted in extracting resources. The dominant political forces in these different communities will put different pressures on the managers of the wildlands in their regions.

Even in more diverse communities, people who have lived there longer and are allied with the historical economies of the past will claim a greater voice in how the past should be projected into the future. The local politics puts a demand on land managers to respond. And respond they have, though in ways that have made them unpopular with almost all of their constituencies.

*Chapter Three*

# Ecosystem Management and Beyond

In a memo in 1992, the then Chief of the Forest Service, Dale Robertson declared ecosystem management as the new strategy that would guide future management decisions about the national forests.[1] He defined ecosystem management as

> . . . the use of an ecological approach to achieve multiple-use management of the national forests and grasslands by blending the needs of people and environmental values in such a way that the national forests and grasslands represent diverse, healthy, productive, sustainable ecosystem.

Ecosystem management has been greeted with skepticism from all sides, from representatives of the timber industry to the whole spectrum of environmentalists and preservationists. What was this new "paradigm" the Forest Service was embracing, and why has it generated so much controversy? Part of the reason is the very generality and ambiguity of its definition.

I have argued that trying to use multiple-use management of our federal wildlands is part of the problem, and there it is squarely embedded in this "new" approach which again will somehow "blend" the needs of people and environmental values. Just how and with what priorities is not clear. The scientific standards to be used are contained in the words, "healthy," "productive," and "sustainable." It sounds like another version of the agency trying to be all things to all people while being sensitive to whatever the political climate happens to be at the time. Even the U.S. Accounting Office got into the act by stating that "if ecosystem management is to move forward, it must advance beyond unclear priorities and broad principles."[2]

Some of the criticisms even within the forestry academic community were openly hostile, including accusations that ecosystem management was just a "bandwagon," and that foresters should not jump aboard. It was "anti-Western, anti-American, and definitely anti- Christian."[3] Less extreme were the comments of the President of the Society of American Foresters in defending the historical emphasis on timber production as well as the role of plantation forestry and clear cutting as having a place in ecosystem management:

> Timber provides economic benefits that dwarf those produced by recreation and other noncommodity uses of the forest. . . . These facts are often overlooked as the different interest groups develop their own principles, criteria, and indicators under the auspices of ecological forestry.[4]

It would be unfair to suggest that these are representative of the views in the forestry profession. A review of articles in forestry journals did reveal a defensiveness toward the notion that ecosystem management suggests that foresters do not always know what is best. Challenging the decisions of when, where, and how trees should be cut and managed is a part of ecosystem management that is an affront to many corporate or academic foresters.

Consider the debates swirling about protecting old-growth trees in the national forests, particularly those outside designated wilderness areas. Well into the 1980s, virtually nobody in the forestry profession even considered that large areas of old-growth outside of national parks might be worth keeping.[5] The public disagrees, irrationally in the eyes of some foresters who argue that forests cannot be preserved indefinitely. In their opinion, today's "cathedral groves" will become "over mature," unhealthy, and die off.[6]

The suggestion that allowing old-growth to die off makes for an unhealthy forest again demonstrates the need for foresters to manage. The tradition of Gifford Pinchot still dominates the profession because many foresters continue to believe that healthy ecosystems means scientific management. It is a mind-set many find hard to shake off. Wood is needed, and if ecosystem management decreases the timber harvest, that is a matter of concern. And in general, ecosystem management will decrease timber harvesting from large forest landscapes.[7] Can you even do this? Some worry that managing for native species and ecosystems while producing a timber harvest is an assumption of compatibility that has not been proven.[8]

Many foresters don't feel threatened by ecosystem management. Instead, the previous era of deciding where to clear-cut has been called "black and white" forestry. They embrace the concept and the movement toward a more biologically sensitive forestry.[9] One presumed movement in that direction is in what has been called "New Forestry." New Forestry is presented as an alternative to clear cutting since, although it involves cutting about 90 percent of the trees in a stand, it leaves some snags and debris in with the remaining trees. New Forestry promises much, a cut almost as high as with clear cutting yet without significant damage to the ecosystem. You get most of your cake, and almost a free lunch. Needless to say skeptics and critics abound.[10]

New Forestry exposes the dilemma ecosystem management faces. Its goal is to integrate the needs of humans and ecosystems, but in the eyes of its critics it remains highly anthropocentric. Humans and their needs get first priority, and ecosystems are managed to try and avoid as much damage as possible. The very assumptions underlying ecosystem management may be false, that nature can be used and controlled as we see fit.

Critics charge that to consider ecosystem management as the solution is to set out on a destructive course that is destined to fail.[11] If ecosystem management is to work, the emphasis should be on protecting the integrity of native ecosystems. Commodity production is tolerated only if it does not interfere with this long-term goal. The interpretation of ecosystem management is quite different, with environmentalists hearing "ecosystem" and foresters "management."[12]

This difference in emphasis may explain why foresters accept this new paradigm more readily than many environmentalists do. For the "every day is Earth Day" foresters it is what they have been doing all along—managing for ecosystems. Their faith in their ability to manage simultaneously for commodity production and ecological integrity remains unshaken. Management remains the basic component in ecosystem management, and that is what they have been educated to do, and do with a passion. For many foresters the mandate of multiple-use remains strong, and a man-centered ecosystem defines the relationship with America's wildlands.[13]

One change that ecosystem management forces, regardless of whose interpretation, is a shift in scale. Ecosystems are defined by nature, and operate on nature's scale, which usually is quite different from what land managers are used to in managing for commodity production. The scale becomes much larger. Managing landscapes for grizzly bears and other wildlife is not the same as managing 20-, 40-, 100-, or 1,000-acre parcels for timber production.

The problem of managing across conflicting jurisdictions and philosophies, as in the Yellowstone example discussed elsewhere, rears its ugly head again. Ecosystem management requires much cooperation both within and between land management agencies. Even if such cooperation can be assumed, a host of priorities have to be set, and it is not obvious which criteria to use.

## ECOSYSTEM MANAGEMENT AND GAP ANALYSIS

The implementation of ecosystem management requires focusing on landscapes as a whole. Fortunately, there has been a start in that direction with what has become known as Gap analysis. Gap analysis largely is the result of the dogged work of one man, J. Michael Scott, a biologist with the U.S. Fish & Wildlife Service. It all started with birds in Hawaii.

Mike Scott was charged with doing an inventory of the forest birds of the Hawaiian islands. To do this, Scott and his team made Mylar maps not only of the birds and their range, but also of the vegetation and land ownership (private, state, or federal). Following a method made famous by Ian McHarg in *Design With Nature,* they overlaid transparent colored maps on top of each other. They found that many of the highly diverse bird habitats were outside the state or federal protected preserves. If the aim was to protect the most birds, there were definite gaps in doing so.

From Hawaii, Mike Scott went to California, where from 1984 to 1986 he was the project leader for the California-condor recovery project. Here he saw first hand the dilemma and high cost of trying to save a species as it approached extinction. To do so on a species-by-species basis was both a politically controversial and expensive proposition. After the condor experience, Scott moved to Idaho, where he developed what would grow into a nationwide Gap program.[14]

The idea for Gap was straightforward. Instead of waiting for species to approach extinction, try to protect large areas with the most diversity to prevent or minimize the chances of species becoming endangered. Determine how best to protect large ecosystems in order to preserve the most species diversity. Search out those landscapes "richest" in species diversity. Do so by taking advantage of data available from already existing databases and especially from what can be gleaned from satellite imagery.

Instead of doing map overlays by hand, use the rapidly advancing computer technology of geographic information systems. Get as much information on species distribution and habitats, vegetation, and land ownership to find the places with rich biological diversity. Then promote saving those areas and their ecosystems as a top priority. Doing so should decrease the amount of future debates about endangered species by decreasing the numbers of species that will be endangered.

It also will identify gaps, "hotspots", or areas where there is rich species diversity but that are not being protected either because they are privately owned lands, or because the state or federal agencies are managing them for commodity production. The gaps also may fall between different agencies where overlapping areas are being managed at cross- purposes, such as is the case in Yellowstone.

Mike Scott became a man on a mission: discover and protect lands with the most diversity. Shift priorities from lands of low diversity to those with high diversity. Put your emphasis on protection, where it will do the most good. When gaps are found, change management strategies or work to acquire private lands where diversity reigns.

In this information age, getting the data needed for Gap analysis is not as straightforward as it might seem. Finding vegetation maps, for example, and getting them in the computer sounds simple enough. But accurate vegetation data is not that easy to come by, nor has it been computerized in many states. Reliability of the data can pose vexing problems as well. Mike Scott started pursuing these and other problems in doing the first Gap analysis for the states of the Northwest. Then he got a bigger patron, and Gap analysis took off.

Bruce Babbitt, the Secretary of the Interior, seized upon Gap analysis as the analytical tool that could help prevent future endangered species "train wrecks." The Geographic Information System (GIS) based Gap analysis would provide a means of heading off future species crashes by protecting their habitats on the federal wildlands under his jurisdiction. Secretary Babbitt went a step further and established the U.S. Biological Survey, which pulled in scientists, primarily biologists, from the various agencies in the Interior Department and put them in one place to work together. It appeared to be a good plan, and at the center stood GIS-Gap, which would provide the analytical capability to make better and more rational decisions. Such were the hopes anyway.

The U.S. Biological Survey got caught in the maelstrom of politics. Angry over Secretary Babbitt's establishment of the Biological Survey without their blessing, Congress eliminated the agency as a separate entity and merged it in with the U.S. Geological Survey. Funding for Gap was cut, and remained at the mercy of politicians more than of scientists. At the same time it came under increasing criticism by scientists both inside and outside the Biological Survey.

One of the strong points of the GIS-Gap project was that it produced maps that could vary in size from large to an easily distributed postcard size. The choice of colors could dramatically show anyone in a few seconds where the unprotected "hot spots" or gaps were. There was no need to plow through reports that earnestly discussed the range of either plants or wildlife. Your eyes need not glaze over from reading bureaucratic or scientific prose. There didn't have to be long discussions of all the uncertainties inherent in the underlying data upon which much of scientific analysis is based.

The colored Gap maps focused the eye and mind, and often allowed for self-evident conclusions to be drawn. The very effectiveness of the mapped output led to increasing scrutiny of how it was generated. Was it a computerized genie that was bestowing the gift of ecological insights, or a mischievous siren, well meaning, but creating expectations that ultimately could not be met? In reality, it was a bit of both.

The criticisms of Gap derive from the level of ecological data that can be obtained for large areas, whether of vegetation or wildlife habitat.[15] The Gap projects use a coarse data set. The minimum size of an aerial unit is 100 hectares or 247 acres. Plants or wildlife with habitats smaller than 100 hectares simply are not shown. The coarse resolution means that the maps leave out a lot of fine detail. Lots of species are in small habitat patches within the 247 acres, and in many places they make up the majority of species. Also, the primary means of predicting the amount of species is by the type of vegetation present. While some wildlife species are associated with specific types of vegetation, many are not, and more complex models are needed.

I cite these criticisms mainly to give a flavor of the types of issues that can be used to attack the scientific validity of making decisions using the Gap maps. Some of the criticisms are being overcome, while others remain contentious. Inherent in the debate is an ongoing conversation among biologists (sometimes quite heated) about the relative merits of small-scale field-documented research, and larger-scale landscape computer-generated approaches such as Gap, and the balance that should exist between them given the limited research budgets of these land-management agencies.

Some of the criticisms will be and are being addressed as the minimum size becomes smaller, and higher resolution maps improve the accuracy of using Gap for plants and wildlife with smaller habitat requirements. The adequacy of the data and models used undoubtedly will be debated by biologists for a long time, given the limited state of knowledge in many of these areas.

There is an old adage that making decisions based on very limited information is better than making them on no information at all. Gap analysis, irrespective of debates about issues of spatial scale and the drawing of inferences from simple map associations, has focused the debate and set the standard for other kinds of analysis to be measured against.

The results of Gap analysis and the data behind it has been made available to anyone who has access to Internet.[16] Right, wrong, or in between, Gap Analysis results have not been locked away in government file cabinets or on computer disks, but are out where they can be easily accessed. Environmentalists as well as federal scientists can use the same data, coarse or not. This is a major step forward compared with the common practice of government agencies telling the public about the results of their analysis. The public can use the same databases as federal scientists. To get some idea of the radically democratic nature of this approach, imagine that the Department of Defense released on Internet the data and scenarios of what it would do to come to the aid of countries considered vital to U.S. interests instead of classifying them as top secret.

Gap analysis is a step toward using an ecosystems approach to managing our wildlands. It opens up the process by going beyond simply including the public. It provides data needed to make decisions, and allows for scrutiny and debate about both the underlying data and the consequent analysis.

Providing easy access encourages the public to ask questions. Doing so can lead to more and more controversy. The Forest Service and Bureau of Land Management had it easy in the past when their use of science was not questioned. Increasingly, they must respond to outside experts who question their data, science, analysis, and decisions. Surely, an understandable response is to turn inward and defend your position against criticism. The policy of the Gap project of encouraging the wide dissemination of its data provides an all too rare example of openness that the other land management agencies would do well to emulate.

## MANAGING ISLANDS OF WILDERNESS

The management of wilderness areas has often consisted of treating them as islands. There has been a lot of emphasis on how much wilderness there should be, but not on how it should fit into a region. Although wilderness is part of a spatial landscape, managers and agencies have treated wilderness as isolated units. For example, the Forest Service continually has rejected the use of concepts such as buffer zones to protect wilderness.[17] Wilderness cannot survive as an isolated island and yet the strategy of the land management agencies promotes insularity.

The island mentality has pervaded, and continues to do so, the selection and management of wilderness areas. There is a need to link wilderness areas with other public lands. Buffer zones alone would not be enough—corridors are needed to provide linkages between other wilderness areas. We isolate wilderness areas only at the risk of having them fail as dynamic ecosystems.

Imagine building towns or cities with no roads linking them together. We wouldn't consider doing such a thing because we know that isolation diminishes community life. Then why do we set wilderness areas aside in isolation? Ideally there should be an interconnected hierarchy of wilderness areas. Practically this is often impossible. However, wilderness can be better integrated into surrounding public lands. As mentioned elsewhere, the tragedy unfolding in the Yellowstone ecosystem is an obvious example of the island mentality within our agencies.

The push towards ecosystem management in the Forest Service should provide hope that a more enlightened and holistic policy that defines the role of wilderness in larger ecosystems would be discussed and debated. Indeed, ecosystem management should provide a renewed debate about what it means to be wild, and how wild our public lands should be. Instead, amazingly the word "wilderness" hardly ever comes up in these discussions. Plough through journals, agency documents, or agency studies, and you will be hard pressed to find the word "wilderness" mentioned, much less discussions of how existing and future designated wilderness fits into the new philosophy of ecosystem management.

A major reason why wilderness is not part of the debate over how ecosystem management is to be implemented is that it is still largely mired within the multiple use commodity-based concept, although in a softer, gentler version. The emphasis is still on how many commodities can be secured while protecting the ecosystem. A change in emphasis, not necessarily direction. Wilderness does not fit well into such discussions of ecosystem management because, according to the 1964 Act,

> A wilderness, in contrast with those areas where man and his own works dominate the landscape, is hereby recognized as an area where the earth and its community of life are untrammeled by man, where man himself is a visitor who does not remain.

If the impact of women and men is supposed to be negligible compared to natural changes, this sounds like one way of managing ecosystems. Not to discuss it, or include it, seems short-sighted to say the least. The managers' perspective is shown by the comment that few land managers would dispute the statement that "ecosystem management is more about people than anything else."[18] People must be included since, after all, we must benefit from the wildlands, preferably by having them provide us with jobs if we live near them, or using them as a source of low-priced mineral and wood products commodities. Or we need to be controlled as "recreationists," of which too many are "bad."

At one level managers are dictating or reacting to commodity production, and at the other are restricting the number of people who can enter a given area. Despite the term "ecosystem management," land managers and agencies seem to be ignoring some of the more important policy questions. Fortunately, others are stepping in to fill the void, and to raise and try to answer, however imperfectly, the larger questions.

## THE WILDLANDS PROJECT

In 1992 some environmental activists and academics issued what they called "one of the most important documents in conservation history: indeed one of the most important documents in the last five hundred years." It was the North American Wilderness Recovery Project, or what became better known as The Wildlands Project. The project was spearheaded by no ordinary desk-bound environmentalists or academics, but rather by Dave Foreman, one of the founders of Earth First!; Reed Noss, the editor of the academic journal *Conservation Biology*; and Michael Soule, one of the founders of the discipline of Conservation Biology and the author of the premier textbook in the field.[19]

They defined the mission of the Wildlands Project as protecting and restoring the ecological richness and native biodiversity of North America through the establishment of a connected system of reserves. Theirs is a long-term master plan to allow for the recovery of whole ecosystems and landscapes in every region of the country.

The existing wilderness, national parks, and wildlife refuges are not adequate because they are too small, too isolated, and represent too few categories of ecosystems. In their eyes, true wilderness and wilderness-dependent species are in precipitous decline. A large part of the reason for such decline is the establishment of wilderness and parks as places that protect scenery, provide recreation, or create outdoor zoos.

They reject the notion that the primary purpose of wilderness is to provide remote, scenic terrain suitable for backpacking. Rather, the Wildlands Project calls for setting aside reserves to protect wild habitat, biodiversity, ecological integrity, ecological services, and evolutionary processes—or vast interconnected areas of true wilderness. To do so they would reverse the policies of the land-management agencies, and not condone the discussions of how much human impact will be allowed on the public lands. Instead they envisage vast landscapes without roads, dams, motorized vehicles, power lines, overflights, or other artifacts of civilization. They assert that such wilderness is essential to the comprehensive maintenance of biodiversity, and although setting such large areas aside is not the answer to every ecological problem, without it current trends toward biological poverty will only continue.

The suggested implementation of The Wildlands Project centers on the work of Reed Noss, who has proposed building upon the existing protected wilderness and park areas, which would serve as core reserves where biodiversity and ecological processes would dominate. These core reserves would be linked by biological corridors to allow for the dispersal, migration, and interchange between plants and animals.

The core reserves would be surrounded by buffer zones, within which only that human activity which is compatible with the protection of the core reserves and corridors would be allowed. The buffers would be managed to restore ecological health and revival of threatened species. Outside the buffer zones normal agricultural, industrial, and urban activities would continue.

The implementation of The Wildlands Project would take place over many decades. Areas already wild would be protected immediately, and those identified as degraded would be restored. Clearly, what distinguishes this project from the various shades of ecosystem management proposed by the agencies is that it is biocentric, making the assumption that all forms of life are equally valuable. Humans are not worth more than other animals; often they are worth less—a radical notion to many people, especially land managers, since nature's primary role can no longer be regarded as the provision of resources for society.

Much of the attention and initial criticism of this project have occurred because it dares to challenge and present a land-management strategy that turns upside down the fundamental assumption that because humans are supreme all lands should be managed for their benefit. The needs of nonhuman species take precedence over human needs and desires.

The Wildland Project does not shrink from thinking big. Although it is long range in scope and promotes an incremental strategy, the expectations are bold. Instead of having less than five percent of the landscape as wilderness, it foresees about one-half of each of the lower 48 states consisting of protected wilderness. This projection applies not only to the western states but to the central and eastern states as well. Other estimates range considerably higher, not lower. Again, these are not just wide-eyed environmental extremists motivated by intuition, beliefs, and emotions, but major figures in ecology and conservation biology who are trying to establish the logic of their position in hypothesis testing and scientific rationalism. The preliminary estimates of how large areas should be is based on the needs of large animals and natural regimes. That it is readily accepted that Yellowstone is too small illustrates the size requirements.

The strategy of protecting large core and buffer areas makes much of the human activity in these areas incompatible with the goals of The Wildlands Project. As Dave Foreman put it:

> We cannot allow ourselves to be boxed in by the "reality" of current multiple use management. Commercial livestock

> grazing on federal and state lands cannot be justified ecologically or economically. Commercial logging, with the possible exception of small pole, post and firewood sales, should be prohibited on the National Forests, although non-commercial thinning of plantation and fire-suppressed stands may speed recovery of natural stands. Mining is an inappropriate use of public lands in virtually all cases. Vehicle use off established roads must be entirely prohibited . . . the majority of dirt and gravel roads on the public lands should be closed quickly.

The vision of truly federal wildlands has made some of the residents of these areas very nervous. The role of humans as secondary to the biological integrity of large regions is hard for many to swallow. The vision of The Wildlands Project is to err on the side of biological protection, which puts it into direct opposition to federal ecosystem managers who may sincerely want to protect federal lands, but prefer to err on the side of maintaining the economic welfare of the people who live in communities intermingled with public lands.

The image of earnest activists and policy-driven conservation biologists wanting to drive local people out of their communities in a form of wilderness ethnic cleansing is quite simplistic. Because it will take many decades to bring about such a system, there will be time to adjust to land changes by a number of strategies ranging from private and public buyouts of more land both by individuals and organizations, to tax-breaks for conservation easements, to increased educational and retraining options to help individuals adjust to a rapidly changing regional economy. Indeed, as my discussion in Chapter 8 shows, such a protective strategy will reap greater economic and biological benefits for the people living in these wildlands regions.

At the heart of an implicit debate between conservation biologists and federal land managers is the concept that biology is a better "bottom line" for making decisions on federal lands than the traditional social and economic criteria because, as a species, humans can adjust better to new and changing conditions than either other animals or ecological systems. But while we can adjust to changes within broad ecological limits, ultimately human cultures cannot endure without healthy biological systems.[20]

Many of the plans being developed by The Wildlands Project have as their basis the protection of large areas for animals that have wide ranges, such as grizzly bears and wolves. If their habitat is not protected and they die off in a particular place, bringing them back will be a much harder task than protecting them, and probably politically less feasible. The Europeanization of American wildlands is an anthem to people who want to keep America wild. Saving patches of habitat for wildflowers or birds is not enough.

Carnivores, the animals at the top of the food chain, are the indicator species that will tell whether or not conservation strategies are successful. These animals require habitat measured not in thousands of acres, but millions of acres. There is no agreement about exact estimates, but when thinking about the habitat of these animals the most realistic conclusion to draw is that large is beautiful is more appropriate than small is beautiful.

To many foreign visitors from a crowded and "geographically limited" Europe, the American West appears to be made up of wide-open spaces with a bountiful network of federal lands and parks. How can there be such crises as disappearing wildlife and threatened ecosystems? Are not the national parks generally large enough to accommodate the needs of most species? Actually they are not. And the recognition of that fact is not just a recent phenomenon. Aldo Leopold warned of this many years ago, when many of today's conservation biologists were toddlers, or perhaps just a glint in their parents' eyes. For he said in 1949:

> The National Parks do not suffice as a means of perpetuating the large carnivores; witness the precarious status of the grizzly bear, and the fact that the park system is already wolfless. . . . The most feasible way to enlarge the area available for wilderness fauna is for the wilder parts of the National Forests, which usually surround the Parks, to function as parks in respect to threatened species.

Nor was he alone. Others argued as well that large wild areas were needed to arrest the decline of the large predators in America.[21] They were largely shouting into the void since even the designation of wilderness some 15 years later was not based on the needs of predators, large or small, songbirds, or colorful flowers, but politics. Even so, the political will to seek and obtain the wilderness designation was directly descended from the work of visionaries like Aldo Leopold and others.

Today the most audacious plan to protect habitat areas is the multi-state proposal of the Alliance for the Wild Rockies. It has proposed the passage of the Northern Rockies Ecosystem Protection Act, which would protect as wilderness about 15 million acres of land in Washington, Idaho, Montana, and Wyoming (Figure 3.1). As shown, the proposal provides for large-scale corridors between protected areas.

The Alliance for the Wild Rockies plan seems eminently logical and builds on the concepts of conservation biology. Natural areas do not follow political boundaries despite the piecemeal designation of wilderness state by state. Setting aside wilderness by using political boundaries becomes a particularly futile exercise if ecological criteria and boundaries are used. Logically sound or not, it steps on a lot

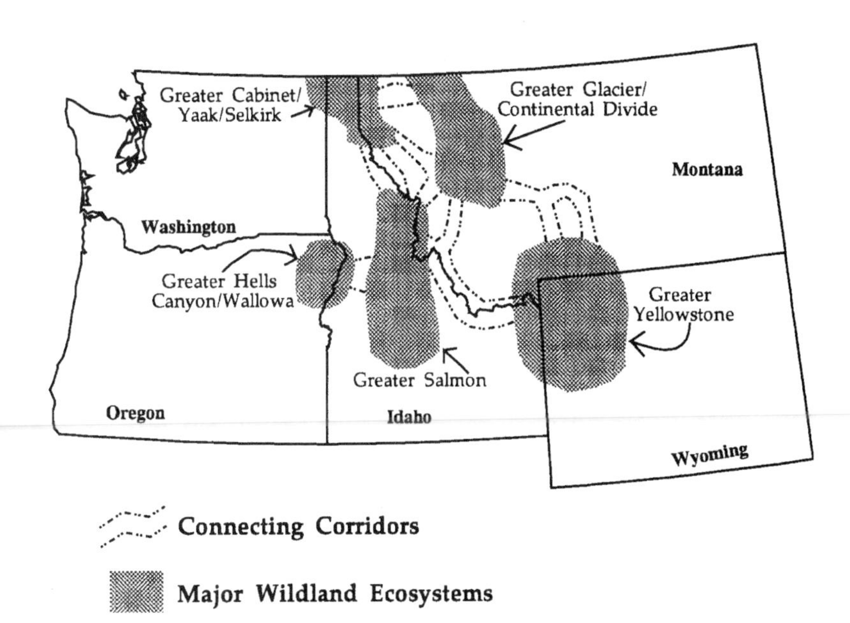


*Figure 3.1* Northern Rockies Bioregion. *Source:* Alliance for the Wild Rockies, 1991.

of political toes and initially was ridiculed as unrealistic by a wide range of Westerners, both environmentalists and politicians.

The Wild Rockies Ecosystem Protection Act has been introduced into Congress repeatedly since 1992 and has gathered an increasing number of sponsors along the way, mostly from states east of the Mississippi River. A number of prominent scientists and an assortment of entertainers signed on as well, bringing much visibility to the proposal. National attention continued to build as prominent environmental statesmen like David Brower and former President Jimmy Carter lent their support and stature. The seed had been planted right in the middle of a region where large-scale integration of currently divided public wildlands was the most possible.

## THE FUTURE OF LARGE WILD ECOSYSTEMS?

The timing of just when a more integrated approach to managing large ecosystems will take place cannot be predicted. It is inevitable, however, and may be much sooner than skeptics are willing to admit. How much longer can public lands belonging to all the people be run as fiefdoms by state congresswomen and senators who have an enormous influence in determining how much wilderness there will be, and how other wildlands will be managed? Imagine what would happen if the same congresswomen and senators wanted to dictate what the nuclear, air, water, or hazardous pollution standards should be for their respective states. Their colleagues in Congress would not go along with it, and yet certain politicians are allowed to dictate land-management policies.

Pollution is a cross-boundary problem and widely recognized as such, requiring regional and national solutions. The protection of ecosystems and wildlands also is becoming recognized as an issue requiring national standards. Even if wildland ecosystems are wholly contained within states, they do not belong to the states. Increasingly using state boundaries as a criterion for public land management will become recognized by an increasingly environmentally sophisticated public as an unsuitable way to keep the American West wild and ecologically self-sufficient.

Certainly there has been a lot of progress in moving beyond the ecological geography of wilderness islands surrounded by heavily harvested commodity storehouses. Wilderness as separate from surrounding lands meant that it was largely ignored by land managers except as a place to manage the impacts of people "recreating" there. The elevation of wilderness areas by some users as places for spiritual cleansing and enlightenment, however temporary, also helps to create a false sense of them as lands apart, somehow different from the larger and surrounding lands.

The movement toward a biologically based management system breaks down such false distinctions. The large-scale geoecological approach shows the arbitrariness of managing wildlands as individual sub-units. Managing wildlands individually promotes the all-too-human tendency of just looking at the consequences of a variety of human actions at a very localized level. The unwillingness or inability to think at various levels has led the land managers to adapt patterns of thought and analysis they are finding very hard to break.

This is quite clear in the ecosystem management debate where, for many scientists and managers, it is a choice between protecting either humans or the wildlands. While professing that humans and nature are interconnected, programs and policies continue to assume they are separate or that one is dominant over the other. Similarly, those who argue for considering all species as equal often have a tendency to either imply or state forcibly that such equality implies separation between humans and other species. This is another false road, for in all of these debates little if no attention has been paid to lessons that can be learned from Native Americans, our predecessors on these lands.

*Chapter Four*

# What About Native Americans and Their Lands?

I live in the Palouse country, a region that is bounded by Indian reservations. To the south near the confluence of the Clearwater and Snake Rivers is the Nez Perce reservation, while to the north is the Coeur D'Alene Indian reservation. Reservations are found not only in this region, but throughout the Northwest and the Rocky Mountain West, and the American West as a whole. Native Americans are part of the landscape, though often an indivisible part.

At my university campus I can walk for weeks on end on land that originally belonged to the Indians, and not see any Native American students. This is understandable, because there simply are not very many at the university. I have had just one Native American student in my classes in over 12 years of teaching. The university would like to increase their numbers on campus.

Chief Joseph of the Nez Perce was captured just 30 miles short of the Canadian border as he was helping lead his tribe away from the pursuing U.S. Army. It was there that he made his now famous surrender statement:

> I am tired of fighting . . . Hear me my chiefs, I am tired; my heart is sick and sad. From where the sun now stands, I will fight no more forever.[1]

The story of Chief Joseph and the Nez Perce was an all-too-familiar one of lies and broken promises, and it followed the chief until the time of his death. Even the final promise to allow him to be buried in the land of his people was broken.

As with other tribes, the Nez Perce had their lands taken from them and were moved to smaller and smaller reservations. The Nez Perce were more fortunate than other tribes that were moved long distances from their homelands to foreign places such as present-day Oklahoma. They were moved across the present border of eastern Oregon to Idaho and were settled on a reservation that steadily decreased in size. The size of their reservation kept shrinking as white settlers kept moving in hoping to find gold in the area. Ethnic cleansing and segregation became the accepted practice of the democratic government in Washington, D.C.

The story of the submission and resettling of the native peoples of North America is not a praiseworthy one. As in the case of most tales of conquest, it has been told from the perspective of the victors, not the vanquished, either in epic histories or popular cowboy movies. Both have contributed to the myths about the American West. What these histories or movies have not shown are the consequences of taking away the rights upon which the Native American societies depended.[2] The destruction of a culture either was ignored or justified under the flag of manifest destiny.

Even on the reservations much of the land and many of the businesses are owned by white non-Indians. The irony for peoples such as the Nez Perce and other tribes is that often their reservations are surrounded by federal wildlands that are owned by the American public at large. The management of these lands for the benefit of the descendants of the people who deprived them of their livelihood must strike many Indians as particularly unjust. The federal lands are grazed, mined, and logged in the name of the "public good," while Indian needs are ignored. Or worse, Indians are stereotyped as lazy and unable to plan and provide for their own needs.

Measured by most objective socioeconomic criteria, present-day

Indian reservations are landscapes of despair and shameful islands of poverty. Their levels of poverty, unemployment, illness, and general societal breakdown match those of any drug- and crime-infested inner-city neighborhood—or even those of one of the poorest Third World countries. Native American tribes are perhaps more isolated and ignored because they are sovereign nations, cultures that stand apart from the mainstream.

The conditions on the reservations cause most residents to leave at some point. At the same time the reservation represents home and a place that Native Americans keep coming back to. Despite all the negative social and economic characteristics of the reservation, a strong cultural commitment to both a place and a people draws individuals back.

The younger generation ventures out into a larger urban-based society and feels pushed to accept the dominant Eurocentric mainstream values of the United States. Many return dissatisfied with what they have found in the larger society. The reservations become islands where tribal culture and religion are protected from the encroachment of the dominant society, and which serve as the last outposts against a debilitating racism that shows few signs of abating.[3]

Returning to ethnic and racial ghettos has not generally been the pattern that American minorities have followed. The history of European immigrants to this country has been one of clustering in cities, living in foul conditions with low incomes and substandard housing, and over the generations, as they progress up the economic ladder, leaving these ghettos to be occupied by other newly arrived groups, such as Asians and Latin Americans, for instance.[4] Whether Irish, Italian, Polish, Japanese, Peruvian, or Thai, the motivation to adopt the American values of individuality and success has created a progressive diffusion of different cultural groups out toward the suburbs, in search of a house and a small plot of land to call their own. Even the stubborn trends of bigotry and discrimination can do no more than slow down this geographical spread outward.

Once "escaped" from the slums and ghettos, these groups have not looked back. In the colonialization process, immigrants had no established ties to the land and neighborhoods they left behind. They are but temporary way stations, though some (such as Cuban exiles in Miami, or El Salvadorians in Los Angeles) may appear as long lasting. Over the course of time they will dissipate and shrink, or make way for others, just as New York's Harlem—which was formerly an enclave of Russian Jews fleeing persecution—became Black and Spanish Harlem.

The American culture washes over and captures a little of the diversity of all the different immigrant groups that comprise our multi-cultural society. Except, apparently, for those who have been residents of these lands the longest, the Native Americans, who still remain a society apart. They exist in a surrealistic limbo-land as sovereign nations that, depending upon historical circumstances, finally "willingly" gave up most or all of their territories to the White Father and his government in Washington, D.C. Their subsequent relationship with the new democracy has ranged from a paternalistic, exploitive, or a masochistic-sadistic one.

What makes the relationship unique is that it derives from treaties made with the native tribes, and the establishment of a trust relationship between them and the United States government. In agreeing to sign treaties to give up their lands, they were guaranteed access to rights to fish and hunt as well as rights to education, health care, and self-government. The guarantees turned out to be of little value to the Indians, or theirs would not be a history of continued exploitation and deprivation.

My intent is not to capsulize the retelling of this ignoble part of our past, nor of our treatment of the original occupants of the present-day American West. Others have done so, although not until fairly recently.[5] To salve consciences it is easier to see those who stand in your way as not worthy. Indians were seen as almost nonhuman savages,; therefore, to replace them with freedom-seeking settlers was a step toward civilization and progress.[6] Nor do I want to romanticize the Native Americans as a people somehow purer and nobler than others. The political struggles for power and control of resources and wealth on Indian reservations and between tribes shows that factions and conflicts exist there as they do anywhere else. Some of these conflicts are due to history, to the voluntary or coercive means of converting Indians, to the determination to "teach" them the advantages of private property economics, and to the general imposition of a different culture.

The various assimilation policies were aimed at getting Indians to fit in, to adapt to the American way. People who press for the benefits of assimilation see no irony in telling Native Americans to "get more American." Native Americans, however, have resisted and maintained a unique cultural identity.

It is not romanticizing Native Americans to say that they have a special relationship with land and nature. Their culture evolved as they lived in specific places and depended upon the land. Certainly they used the land and modified the landscape either by hunting, burning, or building irrigation ditches. The quality of the land the Europeans found on their arrival was the result of thousands of years of Indian settlement. American westward expansion was greatly facilitated by Indian expertise in countless facets of forest and prairie living.[7] That the average American is not aware of this legacy, thereby lacking an appreciation of the roots of our landscape, only serves as another example of the simplistic and distorted view of our past in traditional education from grade school onwards.

Scholars continue to argue the extent to which the Native American use of land is romanticized.[8] But, just as it is romanticizing to assume Indians had no affect on the landscape, it is misleading to state that they dramatically changed the landscape in any way comparable to the way of the Europeans who followed them. For example, the distance between the Indian and the industrial Euro-American view in the Northwest can be typified by the Kalapuya tribe of Oregon, whose view of the new settlers and their plowing of the land rendered the environment "not good."

The Kalapuya viewed themselves as living in symbiosis with the environment, which was destroyed with the arrival of the Euro-Americans. They had lived in the Willamette Valley for several thousands of years, altering the environment most strikingly with the annual use of fire. Nonetheless they also worked within the natural limits of possibilities to create an environment consisting of a complex ecosystem that remained in balance for hundreds of years.[9]

The Indian cultures lived within the limits of their landscape at the same time that they modified their landscapes. Some of this can be attributed to a smaller population and a more limited technology. They did not experience what has been called the tragedy of the commons, a situation in which each person tries to maximize his or her own advantage to the detriment of the rest of the community. The buffalo were only nearly exterminated by the non-Indians.

Lacking such fundamental concepts as private property rights, they managed in a variety of ways to regulate the harvesting of fish and other animals, and to develop agriculture within communal settings.

All of this seems incomprehensible in our post-Soviet world where individualism reigns supreme and collectivism brings images of repression and totalitarian states. The special feel Indians have for the earth and the rooting of their culture in the land explains much. As does their staying in place. Staying in a region for hundreds or thousands of years builds both individual identity and culture. Understanding places takes time. Stories about places build up, and Indian culture is replete with stories—not usually stories of war, but of how humans, animals, and nature relate.[10]

The hunting relationship was one of respect between hunter and hunted. Animals were asked for permission to pursue them for the benefit of the tribe, and were thanked for showing themselves and allowing themselves to be killed. Compared to hunters today, most Indians used as much as they could of whatever they killed. Meat was used for food, skin for clothing, and so on. There was very little waste. Man and other animals lived and shared the natural world together.

Native Americans are held up as having a natural affinity and respect for the world they live in, which is in sharp contrast to the invaders' and colonialists' ethic of superiority, and their need to stand above, subdue, and conquer the environment. This image has been passed down to the present. The most obvious examples are in the New Age groups which try to incorporate part of the Indian culture and spirituality into their own rituals. Needless to say, this gives off an odor of inauthenticity, and Native Americans have not been amused by the modern efforts to expropriate their practices into instant New Age ritual ceremonies.

This cult-like emulation of the environmentally correct practices of Native Americans has a real historical basis. The earlier native peoples were often depicted as purely physical entities, like mountains or rivers that stood as obstacles against the establishment of true civilization. By another view, they were paragons of Rousseau's noble savage who lived in harmony with nature. Though clearly an oversimplification, this latter view is much more accurate than the insulting and despicable portrayals of the Native

Americans as savage, non-agricultural, and barely human. This was a myth, invented to justify their dispossession from the land upon which they depended.[11]

The land on which they depended was quite different from the land of today, which is categorized and often zoned for particular uses, with others excluded. Even more rural land uses are segregated into agricultural, industrial, forestry, grazing and range, ranches and ranchettas, mines and waste pits, skiing resorts, state and national parks, multiple use public lands, wilderness, and so the list goes on. For native peoples such ownership and private rights segregation was unthinkable. Wilderness did not need to be designated, and Indians had no such concept of wilderness. Wilderness is an industrial by-product, protection of the original garden, or as close to it as we can get.

Those were, of course, simpler times. How these native societies would have evolved after being exposed to industrial societies without interference makes for rather interesting speculation. What if these new arrivals, exhausted from their own wars, simply packed up and left, leaving behind only those who would adapt to the Native American ways, and not cajole, threaten, and finally demand that the native people must live on designated reservations. Would the environmental ills that face modern society be similar, or would the strong ties to the land have set them off on a different path, even while acknowledging that the larger outside world could not be ignored? Well, we will never know.

It would be naive to expect that living beyond the environmental bounty would not have occurred in specific places and ways. Certainly, today there are examples where environmental and ecological desecration and damage has been inflicted by Indians themselves. But some rather persuasively make the case that where native peoples have committed environmental "sins," these excesses were generally of recent vintage, and happened under the influence of powerful, imposed non-native economic incentives and value systems.[12]

These excesses generally occur on the poverty-stricken reservations where some resources, especially coal and oil, have been harshly exploited, often benefiting individuals more than tribes. Most recently, some Indian "leaders" have offered reservations as sites for nuclear waste disposal as a way of enabling Indians to "lift themselves up by their bootstraps."

A sad tale, and for which past and current federal government policies share more than their share of responsibility. For even when dispossessing the tribes from their lands and putting them on reservations, they provided assistance, unintended or not, that made living within their traditional cultures even more difficult than intended.

Today, both Indians and non-Indians throughout the West are paying the price for earlier actions. Conflicts between Indians and non-Indians abound, many of which are rooted in the historical disregard for even the minimal rights given to Indians if they would just move onto the reservations. It is as if the negotiators of these treaties presumed that somewhere in the future Indians would vanish from the landscape. They have not, and in places throughout the West they are reasserting rights to a larger protected landscape.

## FROM NATIONAL TO RESERVATION LANDSCAPES

The land upon which all the native tribes now subsist is less than five percent of their original landscape, or about 53 million acres. Even within the reservation, much of the land was opened to homesteading by non-Indians. The landscape of many Indian reservations is dotted with houses and small towns that are predominantly white and non-Indian.

The non-Indian presence is often quite visible. When I go down to pow-wows on the Nez Perce reservation and stop for gas or buy groceries, I am served by non-Indians. The vendors at these pow-wows are predominantly white, although there are very few whites who attend these pow-wows. Although there is nothing wrong with this, it is both unsettling and revealing. Even on reservations and in traditional dance competitions, the modern-day white traders are present. Not much has changed.

Even present-day reservations are not the same size as was guaranteed by the treaties. The Nez Perce reservation shrank in size as the new settlers overran their land, and in "King on the Mountain" fashion proclaimed that they would put it to better use. Towns were built up for these settlers and, as with other tribes, the Nez Perce were pushed off reservation lands. Clearly, much of this was illegal occupancy of the land reserved for Indians, but the federal government generally did nothing, and the state had little interest in the rights of the native populations.

One of the main reasons why reservations such as that of the Nez Perce shrank in size, and why they were moved onto them in the first place, was the search for wealth in the form of ore deposits. Large numbers of Indians in the West were removed from their lands to open up the region for miners. All this was done quite legally, using treaties that moved the Indians to less valuable lands.

Mineral prospectors, led by the infamous Custer, invaded the sacred Black Hills region of the Sioux nations in South Dakota and there they discovered gold. Seekers of gold and glory flooded in and conflicts arose, followed by the Indian war in which Custer met his demise. The Sioux signed treaties and were moved to lands where the scent of gold was not so strong.

The Nez Perce also saw their reservation overrun by gold-seeking enthusiasts and vandals, but they had fought enough. Settlements such as Lewiston, Idaho grew at the expense of Indians such as the Nez Perce. Other places grew and then collapsed, giving rise to the famous resource boom-and-bust cycles that were scattered throughout the West.

Many of these mining towns are being reclaimed by the land. The aftershocks remain in the form of mounds or hills of waste, and as toxics in stream and lake beds. It is hard to visualize the past activities of hundreds or even thousands of people in the remaining decaying buildings that are being overtaken with forest growth in places such as Oreogrande, Idaho. Whether hiking through them or driving by their remnants on highways or back roads, you become aware that they are there as a reminder of a glory-seeking era that long ago died out.

These remnants illustrate the fragile and temporary nature of the "let's get rich quick" economies of the historical West. Few got rich, and those who did mainly lived back on the Eastern seaboard. But dreams and illusions remain strong, as demonstrated in the continual rhetoric of allowing basically free, no-cost prospecting for minerals on public lands and with no royalties to the federal government. Finding precious metals on public lands can lead to personal and, in actual fact, corporate, riches but with little benefit to the public purse via deficit reductions.

Indians have not benefited much from resource development on their reservations. There are some exceptions, such as the Osage tribe in Oklahoma, which benefited from oil deposits on the reservation. Often where such development has taken place, it has been a devil's bargain. The development of coal on the Hopi-Navajo reservation to fuel power plants providing electricity for cities in Southern California and the Southwest created little employment for the tribes and environmental pollution on a grand scale. The pollution from these plants was visible to astronauts as they orbited the earth, and the stacks of these power plants have become known as the "angels of death." they spew out tons of pollution every day and help to decrease visibility in the Golden Triangle of parks, including the Grand Canyon and other surrounding wildlands.[13]

In the growing area of North Idaho, the Coeur d'Alene tribe is still in litigation over the operation of the Bunker Hill Mine, seeking to get compensation for damages and to have the mine cleaned up. The mining of lead and silver has left long-term residents of the area, non-Indian and Indian alike, with the highest blood levels of lead found anywhere in the world. Nearby Lake Coeur d'Alene has toxics at the bottom, for which the best solution seems to be "leave and let be."

Perhaps the biggest disruption of the Indian culture in the Northwest was not putting them on reservations, but the ultimate destruction of the fish stocks upon which many of the tribes depended. Even today as the salmon runs are dying out, they remain a symbol of the Northwest for both Indians and non-Indians. Even more than the eagle, the grizzly bear, or the spotted owl, salmon are imprinted into the culture of the Northwest. Calling on the Endangered Species Act in an effort to save them only shows how far the region has fallen.

It is ludicrous to speak of sustainable development and how it can be achieved, when in the most watered and sparsely populated region of the country, the once wild and numerous salmon return, not in their thousands but in hundreds, or even individually. To suggest that the presence on market shelves of cans of salmon aplenty from Alaska and Scotland somehow makes up for the disappearing salmon of the Northwest is factitious, simplistic, and cruel.

To go further and suggest that pen-grown, barged, and truck-delivered salmon is equivalent to salmon spawned in the wild is to claim that Disney world is the wild. But, so it goes. Academics and others raise such questions, and politicians run with them. The decline of the salmon and efforts to promote their survival parallels in some ways the decline and fall in influence of the Indian tribes that previously depended upon them for their survival.

The story goes back to treaties and making the Indians supplicant and malleable, bending them to the will of the representatives of the Great White Father in Washington, D.C. As a major source of their subsistence, the tribes of the Northwest depended on the yearly runs of fish species such as salmon and steelhead trout that spawned in streams and rivers throughout the region. When coerced into giving up their lands, they made it quite clear that they would not give up their rights to their livelihood, the fish around which their cultures revolved.

They were given emphatic, ironclad guarantees that they could continue to fish where they had fished for generations, even if those lands were no longer theirs. In this way they got a much better deal than Indians in other parts of the country who gave up their traditional hunting and farming habitats when they moved to reservations. This proved to be crucial when the northwest tribes began to assert those rights.

The decline of the fishing stocks and the necessity of calling upon the Endangered Species Act was not because of Indian practices, although some non-Indians did try to make them the scapegoats. Historically, with their regional boundaries and communal societies, the tribes were able to restrict the harvesting of the fish, and create penalties for those who ventured outside accepted practices. The waterways did not belong to any individuals, but the tribes still lived within acceptable bounds, not threatening the resources upon which they depended. The commons from which the tribes drew the fish was not over-harvested or depleted. The imposition of the demands of the Euro-American society that replaced them changed all that.

In just over 100 years, salmon and other fish stocks in the Northwest suffered rapid declines largely from over-harvesting, and the dramatic altering of their environment. From the Pacific Ocean to the waterways of the Columbia Basin, all means at the disposal of the new fishing settlers were put to use supplanting the platforms and gillnets of the Indian tribes. Fishing for subsistence and regional trade was replaced by fishing as a business with little or no restrictions. Whereas the Indian tribes had both unspoken and spoken restraints on the numbers of fish caught and who could catch them, such was not the way of the culture that replaced them.

The rapacious harvesting of fish was accompanied by other rapid changes in the landscape. The era of dam building and land fertilization through government-subsidized irrigation began, and interference with the spawning and migration habits of salmon and other anadromous species was underway. Today, the dams are seen as the major obstacle to the preservation of salmon and other species. The dams were built in such a way that the fish have no means to get around them.

The use of fish ladders to help the salmon get around the dams was initially rejected as unnecessary. Instead, fish hatcheries would "produce" fish that would then swim over the dams, or later on be trucked around the dams and then thrown back in so that they could make their charge to the sea. These strategies have largely failed, and recent evidence indicates that most biologists at the time realized that they would fail. Historian Keith Petersen examined documents and memos that show that federal biologists and engineers knew as early as 1938 the consequences of building the dams—that they were "fish killers." Nonetheless, eight hydroelectric dams were built along the Columbia and Snake Rivers, with the last being completed in 1975.[14]

The dams were needed to meet the demand for cheap power in the region and to help in the nation's bomb-building efforts during the Cold War, as well as turning Lewiston, Idaho into a port. What was the importance of fish relative to the benefits to farmers, grain shippers, bomb builders, and consumers of electricity in the area? Fish over people? Produce cheap electricity for which the Northwest became famous.[15]

Access to cheap electricity and irrigation water has become almost a god-given right to descendants of the settlers of the region.

Meanwhile, the Indians had not forgotten their rights to water. As fish stocks began to diminish, their rights began to be challenged by both commercial and recreational fisherman alike. After all, what rights did they have? It was the settlers who had built the regional economy to what it was. That they had also fostered a variety of problems that would come to a head as the wild and half-wild salmon began to disappear was conveniently ignored. That the surrounding lands were becoming less and less wild was the price of progress.

Today, the people dependent upon the dams charge that they are not solely to blame. What about the over-fishing of the oceans before the salmon turn from the ocean back to the rivers and streams of the Northwest. They are right in that there are a variety of factors that make life more difficult for the fish, and the Indians. Commercial fishing does not regulate itself. The ocean was one big commons, and until governmental restrictions were set, that was reason enough to get as much fish as each fisherman could handle.

Added to the strains on the ability of fish stocks to regenerate themselves were the clear-cutting practices of the Forest Service and private corporations throughout the region, a practice found to be the cause of silted streambeds, thus deteriorating the spawning habitats for salmon, which further fostered their decline. Although dams are widely recognized as the main culprit, a number of other factors contributed to declining fish stocks. Scientists have not been able to reverse the trends. Finger pointing, political and scientific, continues unabated. Earlier voices of doom were ignored, not tolerated, or silenced.

The treaties loomed large as Indian tribes resisted the restrictions on their fishing rights. Although the individual treaties had given them the right to fish, they had not given them any rights to protect or regulate the fish or their surrounding environments. This led to a number of legal disputes with the tribes and the federal government on one side, and states such as Washington on the other.

The federal government had signed the treaties with the tribes and the states were reluctant to allow access to fishing rights at the same time that the supply was decreasing. The courts were asked to decide whether the treaties gave the tribes rights to a specific allocation of fish; if hatchery fish were included in this allocation; and if the right to take fish included the right to protect their allocated fish from environmental degradation.

The tribes were determined under their treaty rights to have 50 percent of the fish, or a sufficient number to provide them with a moderate standard of living. When the verdict came in, they were allocated rights to the hatchery fish as well. They were not given the ultimate right to protect the habitat of the fish. This was to be done in the context of regional planning associations in which the tribes have a voice. Despite this, the reaffirmation of these rights created simmering coals of protest which have yet to die out.

Non-Indian commercial and sport fishermen saw their rights to fish greatly restricted as Indians maintained their rights even as the Endangered Species Act came into play. Even though the declines in stock came about from the practices of non-Indians, tribal members were held to be responsible for the restrictions placed on non-Indians by their very own governments. In local communities tempers flared, and there appeared "Save A Salmon, Spear An Indian" bumper stickers.

Similar controversies arose as Native Americans began to lay a claim to water rights associated with their reservations, many of which had been allocated to farmers to irrigate their lands. Stream and river waters were drawn out as needed to irrigate farms, and the presence of large-scale sprinkler systems is a common sight throughout the more arid portions of the Northwest.

The tribes again turned to the courts and made significant advances, though in many cases the courts had the reserved water rights clearly spelled out and were only required to enforce them. Analogies existed between the subverted rights of blacks in the South and Indian rights in the West in that they both only required a historical conscience, a sense of justice denied, and jurists who regardless of their personal political philosophies would uphold the law. And as they did, Indian rights raised issues many non-Indians in the region would have rather had lie dormant. Particularly if they impacted on their own rights to fish, irrigate, or live more cheaply than they could elsewhere.

## NATIVE AMERICAN–FEDERAL AGENCY RELATIONS

There is an irony in that much of the region that the Indians want to see protected is made up of their former lands, and most of those lands are not in private hands but are now federal lands managed by a number of different agencies, such as the Forest Service, which has most of the "official" wilderness lands; the U.S. Army Corps of Engineers, which builds the dams; and the Bureau of Reclamation, which provides the irrigation waters. The focus of these agencies has been in one way or another to promote development, specifically non-Indian-benefiting development. They have not been much concerned with helping the Indians develop, unlike the Bureau of Indian Affairs, which by all rational assessments has largely failed to help much.

Although the Bureau of Indian Affairs has come in for a lot of criticism over the years—much of it justified—paradoxically to simply cut the bureau's budget drastically or to eliminate it would only make the Indian dilemma worse. Indian reservations remain by and large islands unconnected with the larger society. To simply take the poorest areas of the country, cut them off even more, and say "fend for yourselves" is a cruel solution with predictable outcomes.[16]

As the agency responsible for implementing the trust relationships with the tribes, the Bureau of Indian Affairs has historically played a very paternalistic role. The earlier view is quite evident in the 1900 *Census of Agriculture*:

> Two prominent factors in the development of the Indian are the Indian agent and the government schools. The position of the Indian agent is one of great responsibility and opportunity, and, if the confidence of the Indians be once gained, his influence over them is very great. Under a wise, judicious, and energetic administration, progress may be rapid; but on the other hand, a tribe will quickly retrograde if entrusted to the care of an indifferent agent.[17]

This approach to trying to shape the education and lives of Indians has happily disappeared for the most part, though more subtle versions of getting them to learn from how things are run in the non-Indian society remain. For example, a more recent article discussing Indian land management opened as follows:

> Agricultural success stories are the exception on Indian reservations, where potentially productive land often sits idle, growing weeds and eroding. Despite abundant natural resources, including land, timber, wildlife, and energy, they remain among the most impoverished of Americans.[18]

The implication is clear. Indians are not managing lands to their best and highest economic uses. The Bureau of Indian Affairs and government regulations are held responsible by the authors. The agencies' trust authority over Indian lands makes it difficult to put lands to their highest use by renting them out. Both individual and tribal trust land is subject to agency oversight. Also there are multiple owners and getting agreement between them can be a big problem.

However well meaning, any recommendations based on this logic leads to market and economic dictates as ruling supreme. Become more like us, and life will be better. Indians must have a hard time being polite when they are offered such advice, when looking outside of the reservation they see the environmental devastation and species decline that has occurred as a result of following that advice. Also, the more subtle message of "cast off your past culture," intended or not, must be hard to swallow.

It is not that Indians do not want to develop. Their way may simply be different, and perhaps we should learn from them. Charles Wilkinson, who has done some of the classic work on Indian law, put it well:

> Part of the Indian culture that may, in certain instances, hinder certain types of economic development is the close affinity and respect Indians have for land and nature. Many Indians are reluctant to develop fully the vast natural resources on their reservations because of the adverse impact such development would have on the environment and natural beauty of the land.[19]

In the past most of the management of reservation resources was approved and directed by the Bureau of Indian Affairs. Today the tribes are asserting their power to make more and more of those decisions. For example, the forest lands on the reservations are held in trust by the United States. Indians in the past were limited to stating objectives and putting plans into place. They did not develop the plans. All of this has changed; they are now involved in all phases of the management decisions even if non-Indians are hired to direct the efforts. Forest lands make up about 25 percent (14 million acres) of reservation lands, and about six million acres of that is in commercial forest.

The 1975 Indian Self-Determination Act gives tribes direct management, but the Bureau of Indian Affairs retains oversight responsibility, creating a potential for various local conflicts. The agency now sees its role as technical advisor. The 1990 National Indian Forest Resources Management Act makes the role of the tribes even clearer. Whereas in the past the tribes normally just signed off on timber sales, now the tribal councils make those decisions. Still, much of the advice they get is based on the traditional non-Indian multiple use management in which most foresters are trained.[20] Can we expect non-Indians to manage in Indian ways?

It would be wrong to leave the impression that non-Indians did not recognize the cultural differences and values of Indians that would lead them to practice a more nature and ecosystem management style forestry before it became a buzzword. Back in the 1930s, the Commissioner of Indian Affairs, John Collier, praised the Native American view of land management:

> Tribes supposedly apathetic if not sullenly resentful . . . have stepped to the forefront as conservators, creators of great cattle herds which do not overgraze, the operators of cooperative enterprise of the most modern types. And in their political self-government these tribes have become models, deserving study by the white counties or States. . . . The conservation of water, soil, herbage, and fauna is of national importance, . . . Here the Indians have shown the way.[21]

The tribal forest of the Menominee Indians is an example of a successful forest. After 140 years of management it contains about the same number of board feet of timber as when they began harvesting the forest, and the trees have about the same average diameter. According to their forest manager, Marshall Pecore, a major reason for their success was their refusal to westernize through the development of individually owned property. Still today, the Menominee collectively own their tribal lands.

Although tribal members hold different and conflicting opinions about the effect of resource development on their society, there is a broad consensus that development must not be an end in itself. Toward that end they have practiced sustained-yield forestry. They realized that to survive as a culture, they would have to preserve the forest while they made their living from it.

Over 80 percent of the forest is divided into 109 compartments, to which the foresters return every fifteen years to select the trees to be cut. Then comes the big difference between their practices and those of the U.S. Forest Service and private forest companies, both of which want to cut the best and the biggest trees. That is why we have so little old growth left. The Menominee's rule is to cut the worst and leave the rest. They deliberately pick out trees that are less likely to survive, and allow the healthy trees to keep growing. That is why they have a forest that is similar to the one they had when they started, and not the tree-farms that dot the corporate landscape, which the Forest Service sought to emulate. The Menominee took a different path, and they have a healthier forest.

When writer Wendell Berry visited their reservation, he found that everybody he talked to urged him to understand that the forest is the basis of their culture, and that their goal has always been an old-growth, healthy, and productive forest that is home to them and its wild inhabitants. In Berry's opinion, their "forest economy was as successful as it is because it is not understood primarily as an economy."[22]

There are other tribes, such as the Yakima tribe that have also been practicing selective cutting. The Winnebago tribe has begun reclaiming lands that previously had been leased for agriculture. They wanted to use a holistic approach to reconvert the lands to a combination of forest, agriculture, and agroforestry lands according to plans the tribe laid out.[23] There are encouraging signs that the Indians are starting to do it their own way, focusing on the needs of the seventh generation rather than limiting their planning to the short-term framework that non-Indian "experts" so often espouse.

The new freedom being given to the Indians has its limits, however. Many reservation lands contain Indians and non-Indians alike. It would seem fair to allow tribes not only to decide how they will use the lands, but how they will zone them as well. Here the government and the U.S. Supreme Court has been less forthcoming in allowing them to plan within their reservations. One case involving the Yakima tribe concerned their ability to put zoning restrictions on non-Indian lands within the reservation. The court ruling restricted their ability to do so, thereby continuing an ongoing conflict between tribes and Washington State. The Supreme Court itself has a checkered past since it had set precedents that validated the taking away of lands Indians had occupied as long or longer than had European tribes who settled in Europe. Many now consider this as rationalizing theft and, in stronger terms, the cultural and political genocide of the Native American.[24]

Given that any Supreme Court is not obligated to follow what previous courts have done, why wouldn't they be more willing to allow Indians to manage their own affairs? After all, slavery, racial discrimination, enforced segregation, and other questionable matters that have been upheld by various courts have been overturned by others. Could it be that a fear of allowing Indians to manage too much of their own affairs influenced the court? Perhaps the tribes would ask for contested and other lands back, not accepting monetary payments even if those payments ran into the millions. What court is going to turn millions of acres back to Indians?

We are in a period when Native Americans can assert some of their rights and are allowed to make some of their own decisions. Whether this trend continues or the old custom of ignoring Native Americans makes a comeback remains to be seen.

The lands that often border or are in close proximity to Indian lands are our wildlands. Given a more enlightened attitude toward Native Americans, have they been consulted on decisions affecting their lands? In some ways yes, and in others no.

One of the more egregious instances where they were not included was in the Forest Service's Ecosystem Management Assessment Team which followed on the heels of the Presidential Timber Summit to consider issues and consequences of managing the federal forests of western Oregon and Washington. The intergovernmental team failed to include tribal fish and wildlife managers or to comply with the appropriate Indian treaty rights. They also never considered whether the strategies for protecting fish and wildlife would be consistent with tribal strategies.[25] This was a very big oversight, occurring just as Indians are in an initial period of asserting their rights, and with federal managers claiming they are now more sensitive and receptive to working more closely with the tribes.

The history of state and land management of public wildlands works against them. When it came to co-managing joint resources or wildlife, the relationship was one of "father knows best." Indian input either was minimal or less likely to be heeded. Not all the blame can be put at the feet of the land managers. Indians did not have well trained and educated managers among their own people. When they did, as often or not they were trained in the white man's way of managing for productivity. Taking a critical look at how Indians had managed the land and resources for thousands of years was and is not part of the curriculum in the natural resource colleges of the West. If they objected, the typical response was that Indians simply did not understand how wildlands should be managed.

Perhaps, finally, we are ready to admit that there is much we can learn from Native Americans about how we can live with the wilderness that still remains in the American West. But before we do that we need to address those who would sell off or privatize the remaining public wildlands, and encourage Indians and the rest of us to put our wildlands to their highest uses. We need to consider whether economic rationality or public values is a better guide to how we should "manage" wilderness.

## Chapter Five

# Why Not Sell Off America's Wildlands?

Federal land managers have come under attack by a group of economists, political scientists, and others who believe that by selling off the public lands and allowing the free market to operate the land would be put to its best use, resulting in more wilderness. The rationale is that the problems created by public management—such as its below-cost timber sales, decline in habitat for fish and wildlife, and silting of streams—can be fixed by changing the ownership of these wildlands. These arguments pop up again and again. They are based on an inherent faith that the market is more efficient than public management.

The "new" resource economists who were in favor during the Reagan administration were especially vocal in arguing for private management of the public lands. However, it was economist Milton Friedman who started the privatization debate in his 1968 book *Capitalism and Freedom* by contending that public lands such as national parks and wilderness primarily benefit the well-to-do, but are unfairly subsidized by those who never use the lands. He argued then, as others continue to argue now, that such subsidies should be eliminated, and the users forced to pay the market price to use these public lands. By awarding use of the land to the highest bidder, the theory goes, land is allocated to its most efficient use.

The demand for wilderness would be reflected in the price people were willing to pay for it, or the price others were not willing to pay for it. If the public lands were put on the market, those having commodity values for mining, logging, cattle grazing, or recreational uses for which a price could be charged would be bought up and put to their appropriate highest use. Presumably, much of the public lands would not have commodity values since there would be no federal subsidies for mining, logging, cattle ranching, road building, or activities of any kind. The zero subsidies would result in individuals and companies making their bids based on what profits they thought they could make from the use of the land. If profits could not be made, the land would not be sold, but left in a de facto wilderness state. How much would be in this state is not known.

The "sell off the lands" argument does not take into consideration that the lands—whether "needed" or not—may be bought as a hedge against future needs. The lands also may be purchased on the assumption that political pressure can be later used to institute government subsidies for the extraction of commodities from the former public lands. The past and present use of subsidies to produce commodities ranging from corn, peanuts, or sugar on private lands is often overlooked in the privatization arguments.

A look at history and the role of the market in providing any substantial wilderness or even parklands suggests that little or none would be provided. From Maine to Texas there are few private or public wildlands. Texas has very little forested lands to manage. Maine has one of the smallest percentages of its lands in public ownership, but vast areas of forests. Maine is a clear illustration that private companies do not create wilderness areas or parks. Yet there is little disagreement that the amount of public lands in the Northeast falls far short of demand. If public lands had not been set aside in the past, or wilderness designated, there simply would be none worth mentioning today. Wilderness is not something the market provides. Nonetheless, there is clearly a demand for wilderness, whether provided by the market or not.

The pricing of goods on public lands is at the heart of the argument for privatization. Arguments in favor of private over public land management are based on an intrinsic faith in the workings of the price system. The price system legitimately provides a whole host of products, but unfortunately fails to provide goods such as wilderness or environmental protection efficiently.

For years, economists ignored the value of clean air and water because they didn't have a "price." The tragic consequence has been the production of a horrendously "inefficient" amount of pollution that can be reduced only at enormous cost. The price system works no better in preserving wilderness with all of its inherent values. For example, a private owner may put land to its most economically advantageous use, trading off between harvesting timber, mining, or collecting grazing fees. But what about wildlife that depend on wildland ecosystems? Wildlife has value as a public good because we get both direct and indirect benefits from its existence and/or the knowledge of its existence, but it has no established price.

We value the eagle, the grizzly, and the wolf in the wilds of the West whether we ever see one, or even want to. Only a few would pay money to try to track and kill a grizzly. Rather we gain satisfaction from knowing that it is a part of America's wildness and a small part of us. If we won't pay, there is no market benefit to the private landowner of the nation's heritage. And so other land uses such as logging and mining prevail. From the standpoint of the price system, if these practices reduce wildlife habitat, so be it. On private lands, we as a society have so far been unwilling to pay the price to stop that loss. On our public lands, we increasingly want to prevent the losses associated with losing their wildness.

Privatization of public lands raises the issue of corporations being socially and ecologically responsible in terms of preserving and maintaining habitats. There are companies that demonstrate such practices, trying to maintain wildlife habitat by adjusting logging and other extractive practices. Others try to convey the image of a corporate social and ecological conscience and responsibility through public relations and campaign ads. This poses a dilemma for privatization advocates since they argue that companies will be efficient, implying that at the same time they will be responsible.

But is there such a thing as corporate responsibility. Milton Friedman, the patron saint of the "new resource economists," who, after having written *Capitalism and Freedom* and other works was awarded a Nobel Prize, argues there is not. At least that was his philosophy when I was a student in his graduate microeconomics class back in the 1970s. The only responsibility of the corporation is to make profits for its stockholders. Others have made the same argument. Companies taking social or ecological criteria into account are more likely to be susceptible to corporate takeovers and restructuring. Maximizing profits and doing what it takes to gain a competitive advantage are the guiding principles of corporations.

The evidence from the American West is that timber companies have followed this advice. Again with some exceptions, timber companies have demonstrated that trees can be harvested more profitably by abandoning environmental constraints and by cutting whole sides of mountains without considering wildlife or their habitat.[1] The dictates of the competitive market regularly force decisions that are detrimental to both the forests and the economic viability of a forested region.

Take, for example, the conflict over the harvesting of old-growth trees. There has been an acrimonious debate over the harvesting of old-growth trees, partly because they are the habitat of an endangered species—the spotted owl. Would private companies have any incentive to have old-growth timber on their lands if there were no laws requiring them to protect endangered species? Would the magnificent redwoods and sequoias be left standing under private management? Would trees be allowed to get old? No, both history and economics work against the preservation of old-growth trees.

The history of private logging has been of cutting down all "good trees." The only significant old-growth forests are on public lands. Economists have also shown that, for private companies, managing for old-growth stands will almost never be justified.[2] Unless the cost of harvesting trees is extremely high, the economic rotation age will be reached long before a stand reaches old age. If companies buy land that has old-growth on it, it will shortly be harvested. The costs of waiting for the old-growth forest to emerge makes this an inefficient option on private industrial forests. If old-growth forests are to be preserved or even allowed to reach that stage, they will need to be under public management. Otherwise, the only old-growth will be in inaccessible places, and only for as long as it is not economically and technologically feasible to harvest these remaining stands. If old-growth is necessary for biodiversity or as remnants of the wild, public decisions are needed to preserve it.

A timber company can be a good target for a hostile takeover because of the paper value of its assets—the uncut timber. The more timber (particularly large old-growth trees such as redwoods and Douglas fir), the greater the value of the company's capital. Cut the timber, and the paper value falls, making a takeover less attractive. When such competitive pressures are at work, try arguing for the need to preserve good will and the value of wilderness. Wilderness has no private value unless it can be sold. Private companies do not sell shares in ecosystems, nor do they lobby governments to create wild places.

## BEYOND SIMPLISTIC, FAULTY, OR BAD ASSUMPTIONS IN THE ECONOMIC ANALYSIS OF THE VALUE OF WILDERNESS

There are a number of studies purporting to measure the value of wilderness or how public lands should be used that are evidently wrong. They generally use assumptions or theories that have little value in helping to settle the controversies about how to manage these lands properly. These analyses are based on what has been called "folk economics," in that they seem plausible and are accepted as being valid. The export theory of regional development and the models derived from it are an example. Even though it is recognized that the regional economy is much more complex and that export base theory is seriously flawed, it continues to serve as the basis for much of the economic analysis done by the land management agencies and their private and academic consultants. These models often are used to justify policies such as the impacts of cutting or not cutting timber on public lands, as well as the costs of setting aside wilderness areas.

The inadequacy of the assumptions and the models used needs to be highlighted to expose the inadequate basis upon which decisions about our wildlands are being made. Let me be clear. I do not mean to attack individuals or the motives of researchers who believe that what they have done and continue to do is correct or that it is at least better than not doing anything. Modeling is an imperfect and limited representation of reality. But some of these criticisms have been around for at least 10 to 20 years, while more recent ones have appeared in the literature with suggestions on how to improve the current analyses. Surprisingly, some of the legislation passed by Congress has dictated that more attention be paid to improving the assumptions, models, and analysis used in reaching decisions on public lands.[3] Some attempts have been made, but by and large much more needs to be done. Continuing to make decisions about the use of our public wildlands using flawed economic assumptions and analysis will only increase conflicts over the "best" use of these lands.

There are a number of distinguished economists who have argued that either the assumptions used in various models and approaches are wrong or they ignore critical issues. There is a tendency to ignore what does not fit existing models. As a result, they argue that many economic conclusions are incomplete, inaccurate, or incorrect. Much economic analysis abstracts from reality to such an extent that it is not of much use in the public policy arena.[4]

There are some obvious ways that analyses could be improved. Let me indicate some that are of particular relevance to wilderness policy.[5] The tendency of most studies analyzing wilderness and public lands to ignore non-market values such as environmental amenities is at best misleading, and more often than not leads to decisions that are wrong. Indeed, one of the more absurd assumptions is that setting aside wilderness imposes costs in the form of unharvested timber. It is outrageous to state that anything that interferes with timber harvesting should be considered a cost. The costs of logging on ecosystems is simply ignored, or given lip service.

Little consideration is given to the public interest in having wilderness set aside and the wild nature of public lands protected. It explains why the public agencies opposed the setting aside of wilderness lands. When that strategy failed they recommended classifying as wilderness, only "rocks and ice," or lands with little potential for commodity development. The assumption was that federal lands with commodity values should not be considered for wilderness designation.

The public, in whose "public interest" the land agencies say they are working, has said quite clearly that their priority for these lands is to protect them first; logging or other extractive activities will be authorized only if they do not interfere with that priority. The signals are clear. Just because preservation of watersheds, fish and wildlife habitats, and other characteristics that keep our public lands wild are not sold in the marketplace, they are not unimportant. Just the opposite. They are considered the most important attributes of the public lands. Ignoring these values places the emphasis where it should not be. The focus should not be exclusively on the value of timber or other commodities.

## WHAT WILL YOU PAY FOR WILDERNESS?

It is not that economists have not tried to measure the amenity and environmental attributes of our wildlands. They have tried using questionnaires to get estimates of what people would pay to preserve wildlands, wildlife, and scenic vistas. This is a controversial topic because people are not required to pay, but are asked what they would pay.

The controversy has been about both the validity of the technique and the methodologies used. The National Academy of Sciences established a review panel that included Nobel laureates in economics to examine the issues, and they concluded that the use of these contingent valuation techniques was appropriate and the methodology fundamentally sound. They have not been used by the agencies, though there have been academic studies trying to evaluate wilderness and wildlife.

The studies that have asked how much people would pay on a yearly basis to have additional amounts of wilderness areas in their states have resulted in large numbers, in the order of tens of millions, and have shown that most people would be willing to have much more federal wilderness than they currently have or than is proposed by their congressmen and senators.

One study in Utah found that people in that state would pay on average from about $50 to over $90 per person/year each to have from 5 to 30 percent of Utah designated as wilderness by the federal government. This amounts to an estimate of between $27 to $47 million a year. By comparison, the value to ranchers from grazing their cows on all the public lands in Utah ranges from $2 to $10 million dollars per year. An even more recent study in Utah found that supporters of wilderness would pay, on average, at least $200 per year and that the total estimate of willingness to pay was between $100 and $300 million per year. The authors also found that supporters of wilderness were willing to pay more to have wilderness established than opponents were not to have wilderness.[6] The value of wilderness based on what people say they would pay greatly exceeds grazing values on public lands. These differences become much greater because the estimates do not include the value of wilderness in Utah to people in the rest of the nation. Similar results were found in Colorado.[7]

The political debate over the value of wilderness varies widely, with groups representing the extractive industries arguing for no additional wilderness since any more would cripple local and state economies. One economist even argued that putting land in wilderness would lead to an almost 100 percent decline in the economic activity in Utah. The absurdity of that was quickly pointed out by economists and non-economists alike.

At the other end of the spectrum, some environmental groups will argue for putting all of the federal lands into wilderness. The agency recommendations usually run in the range of between 0.5 to 1.5 million acres in additional wilderness. Studies such as those in Utah suggest the economic value of wilderness normally is significantly greater than what agency officials recommend, and much closer to the recommendation of environmental groups. In Utah the agency recommended about 2 million acres and the various environmental groups from 5 to 16 million, with the survey results showing that people in Utah indicated a preference for between 8 to 10 million acres.

The examples above show that the economic value of wilderness, wildlife, and clean air or water is normally put in the context of what people are willing to pay for them. Multiplying values—whether $10, $50, $100, or higher—by millions of people results in a willingness-to-pay figure in the millions or even billions of dollars. Nonetheless, these are often still gross underestimates.

You can ask for the value of something in two ways: what people would pay to have it, and if they had it, what they would have to be paid to give it up. Economist Jack Knetch has looked at the implications of these assumptions closely. The traditional assumption, on which virtually all economic analysis and policy proposals are based, is that for all practical purposes the measures are nearly the same. No matter how you phrase the question, the results are nearly identical.

This is not true, and there is little or no empirical support to back it up. Instead, the evidence shows that people have to be paid much more to lose something like clean air or wilderness—typically up to six times more. If people are asked how much they would pay to set aside wilderness, and that is compared to the value of the timber on those lands, wilderness will not be set aside using the economic criteria if logging brings in more money than people say they would pay to preserve that land. This can lead to making the wrong decision. If asked how much they would have to be paid to not have the wilderness, preserving the lands may well exceed any benefits from logging them.

Most people do not want the public lands and their wildness to disappear. They think about the environment in terms of not having clean air, water, ecosystems, or biodiversity on public lands. They don't want to suffer the loss, or if they do, they want to be somehow compensated for it. We don't talk of manufacturing wilderness. Nature cannot be easily reproduced. We worry, some more than others, about how much we can lose, and at what price. Is sacrificing wildness, habitat, or a species worth environmental degradation or having some jobs? We try to have both to avoid having to make the tradeoff, but constantly we worry about suffering a loss of one kind or another.

The appropriate economic justification of taking away wilderness should be the loss people suffer. This is much clearer when a group of people suffer a loss such as in the *Exxon Valdez* oil spill affair. Both Exxon and those affected by the spill went out and found economists to determine the losses suffered. Evaluation techniques such as contingent evaluation were used by both sides to determine the extent of the losses. Depending on which side they were on, the polluter Exxon or the victims of the pollution, the estimates of the losses calculated by the "experts" varied greatly. But the whole focus was on losses, not on how much people would have paid not to have had the oil spill occur.[8] The price placed on that loss will vary, and as more is lost, the price gets higher. This is exactly where we are today concerning the price or value ascribed to wilderness. It is continuing to rise.

The types and kinds of losses people will take varies. A logger out in the woods is taking a mighty risk. Logging is one of the most hazardous activities there is. Accidents happen, much more frequently to the logger than to those opposing the logging. Yet the logger willingly accepts the risks, even at times bragging about the danger of his chosen profession.

Persons fighting to protect wilderness from the consequences of logging often sue to protect the forests. Though not in physical danger, they feel a higher sense of risk and loss than the logger whose very life may be in danger. They feel as a personal threat the loss of habitat, some species perhaps, and biodiversity. Cindy and Jim sipping wine in Seattle or New York may feel the risks of the losses to their wildlands more acutely than Michael felling trees with his chainsaw does. Michael the logger has a hard time understanding his losses, given his willingness to take risks, while on others he imposes risks and losses that are involuntary and over which they have less control. Taking risks voluntarily is quite different than having them imposed on you.

## DETERMINING THE TIME VALUE OF WILDERNESS

Economics is a discipline with a short memory. It preaches thinking in terms of the present since "in the long run we are all dead."[9] This is why old-growth trees have little value. They are discounted heavily. Using the economists' method of discounting money, a dollar today is worth more than a dollar in the future. The higher the current interest rate the less a dollar in the future is worth. For example, every dollar gotten for a tree fifty years in the future is worth only about five cents today, assuming an interest rate of six percent. If you could get a million dollars today by cutting down trees or $50,000 fifty years from now in discounted dollars, what would you do?

A standard problem with any benefit-cost analysis of public land is this discounting of any benefits over a time period not much greater than one generation, or about twenty-five years. The use of this economic calculus leads toward a preference for cutting trees and eliminating habitats since we are comparing dollars against pennies. The higher the interest rate at any particular time, the less the value of the old-growth. The value of preservation diminishes as the interest rate rises, and approaches zero in any individual's lifetime. It is hard to avoid the assumption that wilderness becomes more and more worthless as time goes on unless this logic is softened or modified somehow.

Contrast the economic view of discounting the future with the seven-generation view of some Native American tribes. The implications of actions taken today are considered in the context of what their affects will be seven generations from now. Will they diminish the quality of life of people at that time? Or consider the arguments advanced for sustaining the biodiversity of wilderness in perpetuity. There is no discounting of the future in economic terms, or assuming that future generations will be better endowed with ingenuity, capital, and technology to solve current problems, whether loss of biodiversity or storage of nuclear waste. However, when economic thinking is limited only to present-day concerns we assume that those who follow us will be able to deal better than we can with nuclear waste or the loss of wildness and biodiversity.

When considering wilderness and its associated biodiversity is it not unrealistic to assume that most people think only of the present generation. Certainly since Thoreau, the self-appointed defenders and crusaders for wildness do not see themselves as engaging in holding actions or lobbying just for themselves. Preservation of the wild is for eternity. People responding to questionnaires do not want to save watersheds, fish, and wildlife for just 25 to 50 years. These are all indications, however imperfect, that in economic jargon the discount rate should be zero, or even negative when considering public wildlands. Indeed, economists have found that people do use lower discount rates for longer periods and for more important issues, and that they consider future losses more important than future gains.[10]

If people did not use negative discount rates, why worry about global climate change thousands of years in the future, or that nuclear plants will have accidents a hundred years from now, or that waste—nuclear or not—will seep into aquifers, or that insects, birds, or large fuzzy animals will disappear, or—fill in with your favorite place or cause. Conventional market discounting will simply ignore the future consequences of much of what our technological society has wrought, ignoring the critics as misguided purveyors of doom and gloom. Instead, there is increasing recognition that conventional discounting is what is misguided.

## DISTRIBUTING THE COSTS OF WILDERNESS PROTECTION FAIRLY

We judge peoples actions by measures such as their honesty and fairness in their dealings with us. We may not like someone, but we will respect them if, despite our differences, we treat each other fairly. That is the image of the rural West, provincial perhaps, but full of good honest folks whose words and actions can be taken on face value. When I moved to Idaho and before the local bank was bought out by a New York bank, many a college student got a loan on a handshake. Auto mechanics have the respect of their customers and the community.[11] Abusive language or body slams by the local police are unheard of. Differences are respected, and a sense of community prevails.

Times, of course, change, as does the civility between people under stress in a changing community. Fairness is a part of the social responsibility that people feel toward each other, and that we count on if we are to live together as people or communities. That is one reason for the romantic view of small-town life, and a fairly accurate one of where I live. The traditional view of economics does not pay attention to fairness. However, people do care that people are treated fairly and that reasonably specific "rules" of acceptable behavior influence their actions and public policy choices.[12]

Cries for fairness rise up from those affected as restrictions are put on timber harvesting and other types of commodity extraction on the public lands. Throughout the West there has been a decline in the number of jobs in extractive industries. The federal government and its representatives get the blame for putting people out of work. Indeed, some of the job loss is from government actions to protect the environmental stability of our wildlands. However, much of the job loss is from productivity increases due to technology and the resulting need for fewer low-skilled workers, the inability of some companies to compete and make a profit, or the depletion of supplies on private lands. Little or no cries of foul play or unfairness are heard as companies put people out of work for these and other reasons.

The shrill cries for fair play selectively aimed at the federal government rise to a crescendo of protest against a War on the West. Job losses and a way of life is being threatened. Hardly a mention is made of the fairness to those other Americans who do not want to pay the price of losing their wild heritage for the sake of providing subsidized jobs that destroy the public wildlands. Providing relatively high-wage jobs to keep extracting commodities from federal lands as part of public works programs is not what most people consider the major justification for setting these lands aside.

The federal government and the land managers are held to a higher standard. The extractive industries have a boom-and-bust history even when the commodities from public lands are sold at very low prices. People in these industries accept and expect cyclical unemployment. Their children also have a history of moving away and of seeking jobs in greener pastures instead of locally as miners, loggers, and ranchers.[13] The job losses have to be seen in the larger perspective of an overall long-term decline in extractive industry employment patterns. Acceptance of these realities in the private sector do not, however, stifle the rancor and anger of employment losses people perceive to be caused by the actions of their federal government.

The job losses caused in part by the government do create sympathy for policies and programs to mitigate some of these losses. Such remedies are seen as the decent thing to do to compensate people for losing their job.[14] Recent federal policies have tried to do just that. Retraining programs were started for displaced loggers to give them skills to help them find a comparable or a better job than the one they lost in the woods. The retraining programs have not focused on producing fast-food restaurant personnel or chambermaids. There are few comparable programs in the private sector. Fairness is considered essential in the public realm, less so in the private sector, and almost not at all in most of conventional economic analysis.

The implications of these criticisms of traditional economic analysis are clear. Much of the economic analysis ignores the most important issues. Whether the results are fair, to whom, how, and where is simply just ignored. Much of the value of public lands is non-economic, but increasingly important to individuals, communities, regions, and the nation's well-being.

The geographic focus of much analysis is limited to the impacts on local communities and regions with little consideration of the larger national interests. The losses to the larger public from the environmental consequences of extractive activities are ignored despite evidence that such losses may far exceed any economic gains to the local communities. If not ignored, such losses are discounted in such a way that the impacts of current extractive activities on future generations are not measurable. Caring about and preserving wilderness for future generations is not entered into the economic analysis.

If economic analysis is going to be used to make decisions about federal lands, it will have to move beyond the narrow focus of the past. Otherwise, focusing on private costs and benefits more often than not will lead to the wrong decisions being made. Decisions need to be made based on the broader social costs and benefits to society that conventional analysis does not take into consideration. Can such an analysis be done, and should it? There are those who argue that it should not and cannot be done. Economic analysis is not appropriate for making decisions about wilderness or other public lands.

## LET THE PUBLIC AND NOT ECONOMIC ANALYSIS DECIDE WHAT IS BEST FOR WILDLANDS

Many of the necessary extensions of economic analysis require putting some kind of price on non-market goods. How much is clean air and water worth? What will people pay to preserve wilderness, wildlife, scenic vistas, and biodiversity? What tradeoffs will people make between one or the other? Which would they pay more for so that it would be preserved or protected?

Many people (up to 50 percent of the population), if asked, can't or won't say how much these "goods" are worth. They may refuse to try to put a value or price on wilderness, or the worth of a day hiking in the wilds. They don't want to make these choices. They simply want these public goods preserved and the environment protected in an *equitable* way.

People who refuse to answer are not being irrational. They are considering such protection as a democratic right, to be provided through public processes, not private markets. They get angry when they are asked to place a price on wilderness or clean air. As voters they have already asked to have them provided through the public process, not markets. We already own 29 percent of the nation that makes up the federal lands. We already have about 100 million acres of land classified as "official" wilderness. Many people don't want to be asked how much they consider an acre of wilderness is worth. These are questions of equity, not efficiency, and depending upon which criterion is considered more important, the outcome will be different.

The economic approach tries to force people to decide how much they value wilderness or other environmental "goods." But no one can be forced to set a price and sell what they consider rightfully theirs. Some property and the rights that go with it are not for sale. For example, in Montana or Wyoming, ranchers have refused to sell the rights to extract coal or oil from their land even when they have been offered money far in excess of their current income.

The dollars that these ranchers are foregoing can at least be estimated. What about the value of ecosystems and biodiversity, concepts which much of the public is only beginning to grasp. How does a person begin to establish a price for something that they consider important, but cannot imagine selling? At what price will the grizzly bear or salmon be allowed to go extinct? Should estimating the value of species be left to biologists and other scientists? to politicians?

The use of efficiency criteria on public lands almost always ignores equity considerations and tries to get a price established either for what the land would sell for, or a value based on what people need to be compensated for a loss. Anyone who has been to a public hearing on a controversial public land or environmental issue knows that the public is generally more concerned with equity than efficiency. They are less impressed with the economic arguments.

Equity is not in the domain of economists or private business. Efficiency can be quantified, equity typically cannot. Determining which criterion is more important may be critical to the outcome of a particular issue. Equity is expressed most directly in the voting booth. But economics often is more than a match for the power of one person, one vote. So, more often than not, what results is inequitable.

When they are asked to set values, people often feel that they are being forced to play a game by a set of rules they don't understand or agree with.[15] Their answers will be used to generate statistical equations relating to the average person within some level of confidence. When asked for the value of wilderness, if the respondent does not answer, her opinion is not counted in the sample. If she feels wilderness has an extremely high value, perhaps infinite, will any extremely high sum given as an answer be thrown out as a sign that she did not cooperate with the interviewer? How does this deny the rights of those who believe that they have to protect ecosystems and the Earth above all else? Should she be required to put a monetary value on non-market policy preference, when she does not know how to answer?

The wilderness purist is horrified about these economic procedures. The ethicist protests that equity and justice is being ignored. How we treat wilderness and our public lands is not simply a matter of economics, but also a matter of conscience. We owe a responsibility to these lands, and some of our emotions about the land rise out of the collective unconscious, yet we are asked to place a price on them in minutes. We are dealing more in issues of individual and collective values and principles. We need to keep these "nobler" decisions separate from more ignoble ones that depend on our willingness to take out our wallets.

Most of us spend our lives switching between the two realms, deciding what is in our personal or family's self-interest, and what is in society's best interest. We set values and prices to satisfy our own self interests, but not society's. American history is littered with attempts to sell off our public lands, putting them on the butcher block, and in so doing tame their wildness. They have all failed. The public refusal to allow this to happen shows that these wildlands appear to be a basic public interest. The public has refused to accept a price—any price—for them.

If someone comes onto our public lands and does harm to them, their managers, or their non-human occupants, we take legal action against them. Protected grizzly bears and transplanted wolves roam within my home region. When someone shoots a bear, a wolf, or one of their cubs, a reward is posted, and if caught the accused person is taken to court. If convicted, the person is fined and sentenced. The bears and the wolves are not for sale.

The Endangered Species Act, which does not allow for economic considerations in whether species should be saved, is the most obvious codification into law of the national conscience that extinction is not for sale. Similarly, degradation of air quality around the originally designated wilderness areas is not allowed. Unless these laws are changed there is no requirement for an economic test of the costs and benefits of keeping species from extinction or maintaining the air in wilderness areas as clean for tomorrow as it is today.

These public preferences, as expressed in our laws that attempt to protect the wild in our lands, have their costs. Laws may have noble purposes and yet inflict the costs associated with halting development, causing the diminishment in the growth of certain types of jobs, and keeping progress and our technological society from marching into these places. How valuable are wild places where there are no roads, motorized vehicles, motels, or even permanent tents or cabins? How much do we want places where animals can run wild, protected from the encroachments of an urbanized outside world? How much is nature left to herself worth to us? Do we want more? How much chipping away at the edges will be allowed? How wild can wild be?

The decisions we make as consumers and citizens can be quite different. As consumers we may go snowboarding or skiing in national forests, while as citizens we oppose establishing more ski resorts in other national forests. We may hate those environmentalists who oppose hunting bears or hunting in general, yet we join with them in opposing mining and logging operations in our national forests that might endanger wildlife habitats. We may drive large four-wheel-drive pickups that pollute the air and destroy forest roads, and yet oppose developments that will add to pollution and congestion in our local communities. We might throw beer cans out of our pickups, but be careful to carry those cans along with our toilet paper out of "our woods."

In one set of our decisions we use our private values and in others our public and community values, including those of our potential future generations. Let me give a specific example. A friend of mine has been associated with the mining industry for over twenty years. Within wide limits, he defends the need for mineral resources and the actions of this industry, often pointing out "that if it can't be grown, it has to be mined." He chastises environmentalists who would lock up more land in wilderness and make changes to the mining laws, saying it is like killing the goose that lays the golden egg. He keeps reminding me that the mining industry provides many of the resources needed to make us comfortable in our houses as well as for manufacturing the computers upon which many of us depend. At the same time he demonstrated and was very vocal publicly about establishing his favorite federal lands as wilderness and preventing mining and other extractive activities from operating in or near them.

The setting of social policy is concerned with meeting the basic needs of citizens, a matter not of efficiency but justice. We set social policy to make our and others' lives better, healthier, more fulfilling, and beautiful. As citizens we try to reach a shared consensus about what is important to us as a nation. If wilderness is important to our conception of what America stands for, we make sure we have wilderness. We do not ask it to pass a benefit-cost analysis as to how much it is worth in the marketplace. Our shared values as a community and a nation matter. They are not just the simple sum of the private choices we make about what is best for us and our family only.

Adam Smith, the father of classical economics and the author of *The Wealth of Nations* (1776) is best known for his image of the "invisible hand." Today, free market theologians and their followers may wear ties with Smith's likeness imprinted on them as they intone his words that the market that works best is the one left to operate unfettered by regulation. Each person working to satisfy his own advantage works to benefit society as a whole. The "invisible hand" of the marketplace works most efficiently.

What is rarely reported are Adam Smith's reflections on social policy. He was a moral philosopher as well as an economist. He argued for the efficiency of the free market in the economic realm. But, in matters of social policy, he said clearly that the public should not listen to people of business, or merchants as they were called in his time. He felt that societal interests and those of business rarely converged. Social policy needed to be decided by citizens acting in behalf of the broader societal interests and not those of the business establishment.

A discussion of how and where economic analysis should be used today would benefit from pondering what Adam Smith had to say. Economic analysis should be used to measure the impacts of various courses of action and their consequences, but not to set larger societal goals because to do so robs citizens of the rights and duties they willingly assume in a democratic society.

Citizens have decided whether issues such as slavery, child labor, safety standards, discrimination, and other forms of activity should be allowed, banned, or regulated. Their efficiency should not be a consideration in decisions to ban them. Efficient slavery was, and still is, slavery. This applies also to our public lands. Citizens can decide whether they want them and whether they should be tamed by man and industry or maintained in a natural and wild state.

Once the broader goals for the public Western landscape are set, economists can apply their models and argue about what the costs are. That is what is happening today. Economic analyses are suggesting that the costs of not protecting our wild landscape are potentially very high. This serves a useful purpose, but even if the findings are what environmentalists and others want to hear, they should be wary of using them as the primary basis for making their arguments.

*Chapter Six*

# How Does the American Public Want Wilderness Managed?

An obvious question is how does the American public want to have wilderness managed? How do they see wilderness within the larger context of managing public lands? Do they trust federal land managers? Such questions also need to be put in a larger perspective. What is the general attitude of the public toward environmental protection? Are attitudes toward the preservation of wilderness in line with more general environmental attitudes?

The American public consistently has voiced strong support for environmental protection.[1] One way of looking at the importance of the landscape of the American West is by considering attitudes toward protecting and using those lands. For example, what are the attitudes of the public toward the management of public lands? What levels of protection would they like to see? How important are non-commodity use designations such as wilderness and new concepts such as ecosystem management? Unfortunately, this needed information is lacking because these questions have not been asked. Public awareness of environmental and resource issues has grown rapidly since the 1960s. Despite the lowered or anti-environmental goals of the Reagan and Bush administrations, the American public continued to voice strong support for environmental protection throughout the 1980s and into the 1990s. Congressional attempts to weaken various environmental laws and regulations generally have failed. The Clinton administration has been more ambiguous. It has promised much and has delivered more than the previous administrations, but not enough to satisfy wilderness and environmental advocates in the West.

In spite of the often-repeated cliché that the environmental movement peaked following the 1970 Earth Day, public concern, as shown in polling data, dropped off only briefly during the 1970s and reached new peaks in the 1980s. The early 1970s decline in public concern for environmental quality has been attributed to the public perception that, since laws were passed, new agencies established, and money allocated, many of the problems were solved. During the 1970s, when asked to choose between environmental protection and job losses, people chose the environment.[2]

During the 1980s, when it became clear that environmental protection was not a priority of the government, polls showed a rise in concern. Similar trends have continued into the 1990s. Concern for the environment has shown remarkable staying power. Common to many surveys has been the increasing support for environmental quality even if it meant tradeoffs that resulted in higher costs or job losses. The increasing emphasis on preservation and protection of environmental quality is seen as part of a shift in many industrialized countries from materialist to post-materialist values.[3]

General public support has extended to maintaining the quality of public lands, establishing wilderness, and slowing or preventing resource extraction from these lands. Despite this general perception of support for wilderness, there have been only limited surveys of its strength or change over time. People want to have it both ways: they want wilderness and environmental protection, but they also want a strong economy driven by jobs.

Unfortunately, while enough evidence is available to examine trends in attitude toward air and water quality, hazardous substances, and toxic wastes, there are only a small number of recent surveys of public opinion on national park and wilderness or public land management in general. One exception is a study that sampled 1500 adults nationwide from 1978 to 1988, asking them "if it came to a simple choice between developing new energy resources and preserving publicly owned wilderness which do you think the nation should choose?" The percent choosing wilderness increased steadily from 19 percent in 1978 to 47 percent in 1988.[4] Since then very few surveys asking people to make policy choices have been done.

On a more localized level, a survey of 600 registered Montana voters found that 42 percent favored setting aside a majority of the six million acres of undeveloped federal lands as wilderness, but opening some for development. An additional 15 percent wanted to set aside all of the land for wilderness.[5] This is in contrast to anecdotal evidence and public impressions that a majority of citizens in the western states want to develop natural resources of the region, including those within wilderness areas.[6]

In a survey of 2,670 persons in wilderness counties, I found that a majority (53 percent) agreed that the presence of wilderness was an important reason in their decision to move to or stay in the area.[7] Given the lack of any prior survey data on this issue, it is hard to make comparative judgments. However, this percentage is higher than expected for several reasons. Most models of migration and regional development assume people move to get higher incomes. If income was the main reason why people moved or stayed, then wilderness should not be given as an important reason for moving or staying. As further evidence, most migrants did not get higher incomes even though most were young, and retirement migration did not play a major role in the population growth of these counties. Yet despite accepting a drop in income, approximately 70 percent felt that their lives were healthier, happier, and more enjoyable.

When they were asked whether wilderness areas are important to their counties, 81 percent agreed that they are. They may have different reasons for why they think so—perhaps because wilderness areas offer solitude or open space, or because they attract tourists. Yet with only 10 percent disagreeing, for whatever reasons (economic, social, cultural, or existential), an overwhelming majority feel that wilderness areas are important to their region.

Setting aside wilderness has been strongly opposed because of the argument that doing so "locks up" resources. This is generally true, but most wilderness areas are the result of compromises which excluded designating areas as wilderness if they are thought to have mineral and energy resources that could be economically extracted.

Whether or not people are aware that mining has been allowed in some wilderness areas, when asked if they would support such a policy, most people would not. When asked if they felt mining was an important management option on public lands, very few thought so. Mining and mineral development has a low priority both on wilderness and public lands in general.

Public lands in the form of national forests are nearly always associated with wilderness areas. People living near these lands use these other public lands about half as much as they use wilderness lands. Wilderness apparently plays a strong role in the lives of people in these counties. This is in sharp contrast to anecdotal evidence gathered from impressionistic bumper stickers seen in the West reading "Wilderness: Land of No Use." Historically, public land managers also have had a tendency to assume that preserved and protected lands are less important than lands that have commodity or recreational uses.

Public preference for management of federal lands in less extractive ways has become clear in several recent surveys. One comparing Oregon residents with a national sample found that in both groups a majority wanted a ban on clear-cutting and greater protection for fish and wildlife habitats. Both groups supported managing forests with a holistic ecosystem approach; preserving natural conditions was considered important even at the risk of losing jobs and revenue. The similarities in the findings between residents of a major timber-producing state and a national sample are striking.[8]

In a survey of people in a 100-county region in Washington, Idaho, Montana, and Oregon, when asked if they were concerned about how federal lands were managed, 92 percent of the respondents said they were. When asked how these federal lands should be managed, the most frequent responses were to manage for water/watershed and ecosystem protection. Only 16 percent cited timber harvesting as the preferred management strategy. This is in contrast to journalistic accounts of this region and the West being dominated in general by persons who want to cut trees, graze cattle, and dig up minerals. Reality is more complex.

When given the choice of either protective or commodity-based management strategies, people overwhelmingly (76 percent) chose protective management strategies. These include recreational uses, protection of water/watershed, preservation of wilderness values, protection of fish/wildlife habitats, protection of endangered species, and protection of ecosystems. The commodity-based strategies (which received 24 percent of the vote) include timber, grazing and ranching, and mineral exploration and extraction. Much of the historical image of the American West is based on the primacy of these timber-harvesting, ranching or extractive industries. Journalistic accounts of the current West as still populated by people holding primarily commodity production values are exaggerated and misleading.[9]

Surveys show an emphasis on protecting the environment rather than promoting development by extracting commodities from the public lands. For example, in Washington State, 57 percent consider themselves to be environmentalists. The number of residents taking some environmental action is increasing. If asked to choose, people opt for "protecting the environment even if that means that some people will lose their jobs and the government will have to spend a great deal more money," instead of "providing more jobs and expanding the economy even if that means some damage to the environment."[10]

I have given results from various surveys because they point to the direction in which attitudes in the American West are evolving—away from supporting commodity production and toward environmental protection of the public lands. This does not mean that people in the West have thrown away their cowboy boots for hiking boots and want a totally transformed West with no tree harvesting or cows on the public range. Not yet anyway. When asked how important harvesting trees, grazing and ranching, and mining are as a use of federal lands, a significant segment of the population still consider them to be important activities.

When asked to list strategies for public land management, harvesting trees varies from about fourth to sixth most important, while ranching and grazing and mining were in the sixth or ninth position. This same ranking holds when people are simply asked how important these commodity strategies are on a scale running from not at all important to extremely important. About 65 percent list timber harvesting as at least somewhat important, while grazing and ranching is considered so by about 55 percent, and mineral exploration and extraction by only about 30 percent.[11]

These survey results are interesting because they are from the interior West where a majority of the people consider themselves to be politically conservative.[12] And yet, while people in the region want some timber harvesting, grazing and ranching, and a small amount of mining, the percentages of those who are in favor of protecting wilderness, watersheds, and fish and wildlife habitats are in the 70th and 80th percentiles. Clearly, even in the conservative West, the majority favor protecting the public lands first, then allowing some tree harvesting and grazing within that overall protective strategy. The emphasis appears to be on good stewardship, with commodity production being allowed only if the ecosystems of the public lands are not degraded.

Although there are some differences between people living in rural and urban places in the Interior West, they normally range from 5 to 10 percentage points. Whereas 76 percent in urban areas thought managing for wilderness values was important, the corresponding percentage was 66 percent in rural places. A difference of 10 percentage points is significant, but not drastic, as the War on the West rhetoric often implies. Water and watershed protection was considered important by 83 percent of rural residents and 81 percent of urban residents, a difference of only two percent.

The differences between rural and urban people was a little higher on the commodity strategies. About 71 percent of rural persons favored some timber harvesting compared to 62 percent of those in urban areas. The differences were less when the subject was grazing and ranching, with 60 percent of people in rural areas favoring these uses compared to 54 percent of the urban respondents. Both groups were almost in complete agreement on mineral exploration and development, with only 32 percent rural and 31 percent urban feeling it is an important use of public lands. These differences indicate a smaller gulf between people in the rural West and other populations than might be expected. Some other differences are rooted in the demographics of the changing West.

Older people in the region are more in favor of extractive strategies, while females and people who moved in within the last 10 years rated the protective strategies higher. Depending on the strategy, the differences are a matter of direction and degree. Age, sex, or migrant status will not change the majority vote for a policy. The next chapter, which discusses the population, demographic, and employment changes, will show why even these differences are in the process of changing.

These surveys suggest that differences are not that great between people in the inner West and elsewhere. Is the West that different in terms of how people want the public lands managed? Are there not differences between regions such as the Northeast, the South, the Midwest, and the West? And what about in the West itself? What about the different Wests—the California led Ecotopia, the arid Southwest, the wet Northwest, and the mountainous inner West? Does this preference for protective strategies cut across these vastly different landscapes?

If surveys about attitudes toward wilderness and other public lands are scarce, breakdowns by region are even more so. This is true as well for regional differences in attitudes toward environmental protection in general, whether for air or water pollution or hazardous wastes. There are only a limited number of surveys that were done to consider regional differences and in those the West is used in its broadest definition, including even California.

These limited surveys suggest that there are not many regional differences on how much environmental protection people favor. Nonetheless, the problem of small sample sizes and large inclusive regions remain. People in the interior West are generally perceived as having less sympathy with environmental regulations and protection than people in urban areas from San Diego to Seattle. My results suggest that those differences may not be as great as imagined.

All surveys are subject to several weaknesses. How people answer can be affected by the amount of information they have, and its accuracy. They can be influenced as well by how the questions are worded. I have been citing only surveys that have no intentional bias built in. They bear no relation to the biased or junk surveys that we get regularly through the mail from political or non-profit organizations that ask us to respond to biased and inflammatory questions and end by asking us to send money to help fight against or promote some cause. They may also come from our political representatives and ask us our opinion on some issue that has already been decided on. The more sophisticated biased surveys target their audiences and may send different questions to various groups of people, depending on how they are expected to answer. These surveys are as worthless as the nonscientific ones that ask people to respond to questions by calling some telephone number. They simply are not representative of the more general population.

Even a properly designed survey can be problematical. People sometimes respond to questions in a way they think they should rather than the way they truly feel. They may say they support something they do not really support. They might not admit to having a prejudice or bias, saying they will not discriminate against someone based on race, ethnicity, or sex when in reality they would.

Protection of the environment appears to have become an issue that a broad segment of the public agrees is important. This seems to be true both for more conservative as well as more liberally oriented voters. When faced with tradeoffs, people prefer environmental protection, or seem to want both a clean environment and a growing economy. When the major environmental legislation was being amended or rewritten in the early 1970s, many people—including academics—thought that there would be a tradeoff between jobs and the environment. It was a variant of the "no free lunch" approach. A cleaner environment would produce a slower-growing economy.

The public seemed to be ahead of the "experts" and politicians of the time because today there has become an increasing recognition that environmental protection can shift the economy in a different direction, generating a demand for different types of jobs. This shift can lead to increased environmental quality and economic growth. This is a significant psychological shift in how the economy is viewed and explains in part why a broad spectrum of people support environmental protection while disagreeing on a myriad of other social and economic policy issues.

Everybody in some way considers themselves in favor of protecting the environment. Just what that means may differ radically from person to person. Persons in the West who profess that they hate the "environmentalists" who are always protesting logging sales will tell you that they themselves are the real environmentalists because they live with and love the land. This is usually greeted with skepticism, outright disbelief, or ridicule by the other side, yet it reflects the environmental attitudes consistently reported by national surveys.

Irrespective of their underlying values, people in the West want to see the long-term protection of the landscape. That is why they rate protection of the forest, watershed, or ecosystem highly. When I first moved to the West I did not understand this, or I did not believe it.

I recall discussions with loggers, managers, and forestry professors, during which they would inevitably say something like, "What a waste it is not to log that land. Don't people understand that leaving trees to rot is such a waste." Usually I kept quiet, mystified because what they were saying went against my own intuition and much of what I was reading, as well as my personal reaction to ugly clear-cuts.

I was more at home with other "experts" who disputed these peoples' claims, and who argued that leaving old and so-called dying trees is actually good for the trees. If you are going to cut, cut the younger, smaller trees that create undesirable crowding.

Perhaps because of my own prejudices, I would be even more surprised when a logger working in the woods would tell me about turning his boss in for cutting trees he wasn't supposed to cut and thereby raping the land. And then there was the academic "expert" who thought all trees on the public lands should be cut before they died and became worthless. Having said that, in the next breath he began complaining about that damned roadcut, and logging on a public forest that dumped dirt in one of his favorite trout fishing streams.

These people do not see themselves as working against the environment. They do not feel that self-styled environmentalists are any more ethical or righteous in dictating how public land should be used than they are. They have a land ethic of place that is just as valid as the next person's.

What different people and groups are arguing about are the consequences of different actions. People with a preservation bias do not want to see more damage done to the land. Persons who want to use the lands for some type of commodity production often sincerely feel that there will be no permanent damage.

An obvious sticking point is that if all these people support the environment, ecosystems, or wilderness, then why do many of these same people support political representatives who vote against many of the issues and policies that they support? Wilderness designation is a very contentious topic out West. How much wilderness or wildness people want on our national forest lands has not been settled in states such as Idaho, and much ire is raised in places such as Utah over potential Bureau of Land Management wilderness. Many local and national politicians in these states oppose any significant wilderness designation in these states.

If wilderness and large-scale ecosystem protection is important, why don't people in the region elect people who represent their views about wildlands? Part of the answer is that the environment is not an issue over which elections are won or lost. Given the wide range of what it means to be an environmentalist, all can profess allegiance to the cause. Whether they view wilderness as land to be used wisely under government supervision in the Pinchot tradition or they adopt Muir's preserve-and-leave-it-be philosophy, all can claim to be advancing the cause.

Whereas opponents cast the terms conservative and liberal in derogatory ways, depending on their own preferences, both often see themselves as the true defenders of nature and the landscape, using and shaping it in the way it was meant to be. Since the 1970s, the American West has been a Republican stronghold, their strongest region in the country.[13] There are variations within the region and states, sometimes based on urban-rural divisions, but overall the trends have been clear.

Do these voting patterns matter? Should they? Will they continue to have impacts on public policy? These are difficult questions to answer. In one sense such questions do not matter. Even though elections have not as yet been decided on environmental issues, neither has the public backed away from the environmental laws and regulations. All indications are that, when the perception of nonenforcement becomes widespread, there will be a backlash.

If being an American means having a West that retains elements of the wild, wilderness values will continue to be embedded in our society. These values will be reflected in how the public demands that our lands be managed. Despite the influence of mass media, the images of cows and cowboys, miners, and loggers are not as strong as those of the eagle, grizzly bear, and wolf in reflecting an independent spirit of survival against the long odds of progress.

People in the West, hikers and hunters alike, depend on images of what the West was, not what it could be in a technological world.

The inner West needs the wild bear and salmon to remain wild itself. A wilderness without the wild becomes a surrounded preserve much like an urban zoo, even if the cage boundaries are unclear.

If the values are broad based, cannot honest women and men examine the facts and come to some agreement? Can they not ask the experts, the scientists, to help in examining the consequences of specific actions?

Science, unfortunately, cannot or will not come to the rescue. Either too little is known, too many scientists subscribe to too many different theories, or the impacts on the wildlife or biological systems are just too complex. For someone like myself who was trained as a chemist, this is hard to comprehend at first. The "hard scientists" often see it as evidence of the lack of "good science."

For both everyday people as well as scientists, it is hard to understand why there is so much disagreement over whether creating open spaces in forests is good for certain kinds of animals. Why cannot fish biologists agree over whether salmon will be "saved" by one course of action or another? Why cannot there be agreement over how much land needs to be wild to provide habitat for wolves, bears, spotted owls, or frogs?

It is because of such disagreement that the issues have become so politicized. Politicians sanctimoniously announce that "Science" is on their side and has proven their positions correct. And since they know, they will legislate what is to be done. They will determine how many board feet of timber will be cut, and how much habitat will be allowed where animals can roam. Smokey the Bear, a cartoon character, is no problem. Unnamed grizzlies roaming the wild West are, and need to be confined within special zones in parks, wilderness, or federal tree farms. If they roam outside their unbounded preserves, they are to be tranquilized or shot dead.

Science in essence matters little; it has been politicized, ignored, or argued to death. Imagine what would happen if politicians decided yearly how dirty the air or water could be in each urban area and told the Environmental Protection Agency what standards to apply where. Most wouldn't think of trying. Yet on federal lands a local congresswoman or senator will offer "advice" on how many trees should be cut, which species should be declared endangered, whether bears should be shot, how many cows a rancher should be allowed to run, and what fees should be charged for grazing. The list goes on and on.

This political micro-management has the federal land managers responding in a very direct way to political pressures. When Congress mandates how much will be cut each year it departs from the usual way it handles most environmental legislation in which it sets general parameters and then leaves the rest to the experts. Are politicians working in the public interest when they become the "experts" on how public lands should be managed? Most of us probably would not think so.

Politicians and others also make use of what is known as "gray science." This is a term that applies to those scientific studies whose results may be suspect but are nevertheless used to make policy decisions. Often they are scientific reports done by or for public agencies. The people preparing them may be academics, consultants, or agency scientists. What distinguishes them from other scientific studies is that they are not peer reviewed or published in reputable journals. Just because the results of a study are not reviewed and published does not mean that they are not sound. However, because many of these results are issued as reports and used to justify policy actions that were planned anyway, they are often suspect.

I have had various scientists in the agencies themselves mention the often suspect conclusions of these "gray science" studies, which have been prepared by academics working in various university departments and colleges. Within the agencies themselves, there is little incentive to publish the research, since the rewards are not the same as in the "publish or perish" academic world.

The use of such studies to justify management decisions has a long history. A good example of how the use of such studies can result in ecological disaster is described by Nancy Langston for the Blue Mountains of Oregon.[14]

The federal managers put a lot of faith in the principle of managing in the public interest. Do they know what that is? The limited survey research on public land management has not been done by or for the managers. They are in a double bind, professing to manage for the long-term public interest, but spending a lot of time reacting to the short-term demands of politicians. The politicians' eyes are fixed on their two-, four-, or six-year election cycle.

The short-term interests of the politicians infect the management policies of the federal agencies whose budgets get tied to how much of the public land commodities they produce, not how well they protect the lands and wildlife.[15] If commodity production is decreased, the agencies must promote other forms of use, by encouraging tourism, charging admission fees, and counting recreational users of public wildlands. Is this in the public interest? Not if, as the surveys indicate, promoting wilderness and land and wildlife protection is the goal of managing the lands.

Should the whims of the public decide how wildlands are managed? Don't they change too often for consistent long-term management practices? Why not stick with the traditional way of telling the public how the lands should be used and managed? Why trust the public? Indeed there are problems with public input, but the alternatives in managing our public lands have been tried, and found wanting.

*Chapter Seven*

# Wilderness and the Communities of the American West

What do we know about the communities of the American West? Not as much as we should. With 84 percent of its population living in metropolitan areas, the West is an urban region, but that is not its popular image. The image and root of Western culture is contained in its small towns. These towns, surrounded by lots of wide-open spaces is what the West is for many people. The real West is not well represented by Los Angeles, Denver, Phoenix, Las Vegas, Salt Lake City, Portland, or Seattle. But mention cities such as Boise, Spokane, Yakima, Bozeman, and Missoula, and it starts to sound more like the West. Consider Riggins, Salmon, Livingston, Republic, Casper . . . and a more authentic West rooted in the landscape of the public lands comes to mind.

Another prominent image of the West is of boom-and-bust cycles historically caused by fluxes in gold or silver mining, and more recently by availability of coal, oil and gas, nuclear reactors, or trees. The people who stay as these trends wash over the land have been called "stickers."[1] These "stickers" cannot be compared with the people they replaced, for the Native Americans lived in the region for thousands of years. Nor can they be compared with ethnic groups in Europe fighting over lands and religion. But for people whose families arrived in the West in the late 1800s or early 1900s there is a certain pride in their identity as Westerners even though 50 or a 100 years is a very short time, in historical or cultural terms, to claim pride of ownership in a region. And yet they do.

Nevertheless, people in the West are a mobile lot. The history of the American West is of continuing migration in and out of places, both large and small. There is the ever-present search for opportunity or flight from some place else. Often the opportunity is based on the fortunes of luck, whether in dredging for gold in the streams and rivers, or playing the slot machines of Las Vegas, Reno, or Laughlin, Nevada.

Consider again Wallace Stegner's insights:

> The principal invention of western American culture is the motel, the principal exhibit of that culture the automobile roadside. The principal western industry is tourism, which is not only mobile but seasonal. Whatever it might want to be, the West is still primarily a series of brief visitations, or a trail to somewhere else.

Ghost towns dry up as resources are drained and people leave, or a Walmart's moves into a neighboring town. Lots of changes large and small, old and new impact the American West. Some are unique to the region, others reflect national and international trends. As a result of these changes, some places grow and prosper and others decline.

The 1980s and 1990s have been a general period of decline for many smaller towns throughout the United States. Especially hard hit have been farming towns. Throughout the Midwest and Great Plains population declines can be traced in red on maps. Some forecasters see an ultimate demise of whole regions as downward-sloping population lines predict a gradual disappearance of these places.[2] But they forget the "stickers" who find ways to stay and survive no matter what.

In the interior West predictions of population losses and economic decline as a result of being on the periphery have not come true. Quite the contrary. Along with the South, the interior West is the fastest growing region of the country. Claims of demise from limiting logging, protecting spotted owls, or even tearing down dams to let the salmon swim free would not change the population trends in the region.

People are moving into the West not in search of jobs in logging, farming, mining, or any of the traditional resource-based jobs in the region, but for a higher quality of life rooted in a sustainable and protected landscape. Evidence of the search for that landscape is rapid in-migration of people who are moving for other than purely economic motives.

The old cliché "you can't eat the scenery" is not true for many areas of the West. Similarly, "wilderness, land of no use" has been a powerful magnet in drawing people into counties surrounded by or containing wilderness. Among the fastest-growing counties in the nation are those adjacent to federally designated wilderness areas.[3] Most of these largely rural counties are in the West and are not within commuting distance of any major metropolitan area.

During the 1960s, wilderness counties had population increases three times greater than other non-metropolitan counties. In the 1970s, they grew at twice the rate of other non-metropolitan counties. Again, in the 1980s, their population increased 24 percent, six times faster than the national average of 4 percent for non-urban counties as a whole, and almost twice as fast as other counties in the non-metropolitan West. These trends have continued into the 1990s. Whatever effects the designation of federal wilderness has had, decreasing the population growth in these areas has not been one of them.

Wilderness areas have not been set aside for recreational purposes, but as places where the impact of people is temporary and unnoticed. Areas around wilderness are not intended to attract people. The same cannot be said about national parks, which have a dual purpose of protection and recreation. This dual purpose has created problems and conflicts for park managers, both inside and outside the park boundaries. Population trends in and around national parks are similar to those around wilderness areas. National park counties have had population increases of 25 percent during the 1960s, 34 percent during the 1970s, and 26 percent in the 1980s (Table 7.1).

My population figures show that people have been moving to areas next to wilderness and national parks consistently. To place

***Table 7.1*** Percent County Population Change

| *Year* | *Metropolitan* | *All Nonmetro* | *Wilderness* | *Park* |
|---|---|---|---|---|
| 1960–1970 | 17.1 | 4.3 | 12.8 | 24.6 |
| 1970–1980 | 10.6 | 14.3 | 31.4 | 34.2 |
| 1980–1990 | 11.6 | 3.9 | 24.0 | 26.0 |

*Source:* Calculated from U.S. Bureau of Census estimates.

these population figures in a larger perspective, I calculated the percentage population growth for the period 1980 to 1990 for areas not classified as either wilderness or national park counties. Since some counties can be classified as both, I also calculated the population figures for these counties as well.

The protected areas of the three states of the Northwest had population increases ranging from 15 to 18 percent, while other counties had increases of 10 percent. The difference is even more dramatic in the mountain states where population increases in wilderness and national park counties range from 17 to 25 percent, compared to only 8 percent in counties without protected public lands. Clearly, the amenities of these protected lands are convincing people to both stay in and move to these areas.

Overall, for the entire West, counties with protected lands grew 25 percent, while those without protected lands grew 18 percent.

These population trends in wilderness counties can be a mixed blessing, especially if one of the reasons people have moved to the counties is to be near and use the wilderness. As population pressures increase, managers find themselves increasingly trying to balance demands for more access to wilderness and national parks with the preservation mandated by the Wilderness Act and the National Parks Organic Act.

## WHY ARE PEOPLE MOVING TO THE WEST AND TOWARD WILDERNESS?

As I said above, the West has a history of mobility. It represents the new, the unknown, the frontier, a place where hope provides visions that other places loaded down with the recent past do not. Other regions carry baggage, such as the witch trials of New England, slavery and its racial after effects emanating from the South and spreading throughout the nation, and the labor strikes of the Midwest. After dispensing with its original inhabitants, the West became and remained a place of opportunity.

The early movement was of pioneers settling a frontier and providing the basis for a new "man" and society. The Western frontier became the means of defining the new American. Then a historian, Frederick Jackson Turner, declared the frontier closed in 1893. The frontier was defined as a line with less than two (white) people per mile. As the line moved westward, and finally disappeared, so according to Turner did the frontier. America became a settled place.

After the West had been settled, the movement of people in the region became similar to that for the United States as a whole—one of people moving from rural areas to cities. Starting with the first United States census in 1790, cities always grew faster than rural areas.[4] Although Thomas Jefferson hoped we would become a country of small farmers who would avoid the corrupting influence of cities, it was not to be. We became a nation of cities, as the percent of the population living in cities increased to 80 percent of the population.

Throughout our history there has often been concern about this mass migration to cities and rural communities have asked "how are we going to keep them down on the farm?" The urban growth in the West has increased dramatically ever since the 1940s when the region grew faster than other parts of the country. The West started to get a greater and greater share of the country's population.[5] Small cities such as Phoenix grew at phenomenal rates, and Los Angeles picked up speed in its drive toward becoming the nation's second largest city.

Before the 1970s rural counties were either losing population or growing more slowly than urban areas. With the 1970s, however, came what was hailed as a rural renaissance during which, for the first time in U.S. history, rural or non-metropolitan areas grew at a faster rate than urban areas. This turnaround came unannounced and unpredicted by the "experts." It was hailed as one of the most significant demographic events of this century, and was declared to be a pervasive new counter-urbanization trend that was destined to profoundly change the geographic structure of the United States.[6]

The 1980s brought a collective sigh of relief to those taken by surprise by the population turnaround of the 1970s. Urban areas were once again growing at a faster rate than rural areas. But not everywhere. New classifications emerged. There were now more remote counties that were categorized as retirement, recreation, manufacturing, farming, energy, mining, or timber counties. This division of rural America into specific types of counties is simplistic, but it provided an explanation for those rural counties that continued to grow. The 1980s also had economic recessions at the beginning and end of the decade. Recessions usually have a greater impact on rural areas.

The rural counties that were not growing included many farming-based counties, and others classified as manufacturing, mining, energy, and timber counties. The counties that continued to grow included those that can be included in a broad-based amenity category with both a desirable physical environment and a relaxed small-town atmosphere. Wilderness counties were placed in this category.

National surveys began to show that, if given a choice, people would prefer to live in small towns.[7] Understandable as that may sound in today's era of constant media exposure to crime and gang violence in large cities, it was not readily understood at the time. I recall a friend giving a paper in 1974 that pointed out this preference for small town living, and the surprised, somewhat hostile audience questioning his results and analysis. The movement of people to rural areas was greeted at first by some prominent Ivy League scholars as a statistical artifact, an error in data compilation by the Bureau of the Census. They were wrong.

Studies also began to show that amenities such as environmental quality and pace of life have become increasingly important in explaining why people move. The apparently sudden preference of people for rural life was a shock to many academics and planners because rural areas were thought to be at a major disadvantage compared to urban areas. Certainly a general movement toward isolated wilderness counties was not expected. It was no surprise that some 1960s dropouts and "return to the land" types might seek out such places, but they were the exception, not the norm.

Theories could not be built around people who were dropping out or detaching themselves from mainstream society. They were not driven by the motivation to maximize their incomes. Earlier studies had argued that economic reasons were paramount in explaining why people move: people move because they want jobs and higher pay. This logic, derived largely from economics, argues that people will do a rough cost-benefit analysis in their heads. What are the costs of moving, both economic and psychological, and will the benefits I receive be greater after I have moved? Will I be better off? In the traditional analysis these benefits are measured in terms of increased income.

The economic model seemed to work well as the country became increasingly urbanized. People moved to cities for jobs and higher incomes. Cities that could provide jobs and were growing would attract migrants. Places that did neither of those things would not attract people, and those that were losing jobs or had high unemployment would lose people to other places.

The acceptance of this almost total focus on the economic rationality of people explains much of the surprise when rural areas began to grow faster than urban areas. Rural areas are not places where entrepreneurs and jobs are. Maybe retired people would move to such places since they no longer had to work for a living, but most retirees who leave their home towns prefer to move to a place that is comfortable, and, which would explain growth in parts of the Sunbelt, but not throughout the nation. The local growth stimulated by the in-migration of retired people was focused around cities where services are available and not in the remote rural areas. Why should retired people move to wilderness and other isolated public land areas where services are remote? They don't move to those kinds of areas.

It became increasingly difficult to explain that the movement out of cities was motivated by a search for higher wages. There was no other major theory to explain the change in pattern. A large number of other explanations were suggested, including the decentralization of many industries, increased mobility because of improvements in transportation and communications, and the growth of recreation and retirement activities, to name a few.[8] These all have a spur of the moment, individualized, ad hoc feel to them. They apply in some instances, but not in others. It is like trying to explain why your favorite team lost the big game. There are either a few big turning points or a lot of little things that did not go right, all of them the result of unexpected occurrences. Different people might not see the reasons for the loss in the same way. Once the economic model appeared to be flawed, other explanations were sought.

Sociologist John Wardwell theorized that differences between rural and urban communities were disappearing, reducing both the economic and social costs of an urban-to-rural move. A harder look was taken at people's preferences. Perhaps if they wanted to live in a small town they might actually have begun to move there. Maybe people had preferred cities and now the distribution of such preferences in the population had shifted in favor of rural areas and small towns. If cities were once considered beautiful, and wilderness threatening and scary, had wilderness now become beautiful, enticing people to move to such places? Questions about societal preferences changing over time are difficult to answer because prior to the 1970s there is a paucity of data on such issues. Such questions simply were not asked, at least not on surveys.

There are several reasons that suggest why the move out of cities and toward rural areas should not have been such a big surprise. For one, the movement out of the cities had already started after World War II with the growth of affordable housing for lower- and middle-income persons in the much-criticized look-alike suburbs with their mass-produced housing. People flocked to them to get their own patch of land and space.

Much of the early movement to rural areas was first attributed to a spillover from metropolitan suburbs. The suburban fringe was simply extending its boundary and becoming more exurban. That was only a part of the explanation. Growth outside of metropolitan areas and near wilderness was far removed from a simple extension of commuting patterns outward to the fringe.

Rural areas are romantically attractive to Americans, most of whom have had little contact with such areas, because of the lifestyles they imply. Also, since the founding of our nation, anti-urbanism has been a part of our intellectual tradition.[9] Today, rather than viewing the city as a place of culture, many see it as an object of fear and a target of disdain. Except for Woody Allen's depictions of New York City, there are very few positive portrayals of cities in our mass media.

Studies done after the mass migration of people to rural areas found that amenities such as environmental quality, pace of life, and low crime rates were the important reasons why people moved.[10] Most of these studies were done in the rural areas of the Midwest, which made the results even more dramatic because it was not a region that was expected to grow. Surprisingly, there were few studies done in the West, and during the 1980s there was little follow-up to see if the preferences for moving to rural areas were similar to those of the 1970s.

In the mid-1980s several colleagues and I did a nationwide investigation of the causes and consequences of migration into wilderness counties. As part of that study, we sent questionnaires to people who had moved into wilderness counties during the past ten years as well as to longer-term residents of these areas. Like the studies done during the 1970s, we wanted to know why people had moved to these counties and how important economic factors were, compared with non-economic factors such as amenities.

Wilderness counties are high amenity counties, and normally it is assumed that high amenity counties grow because of an influx of retirees. We found that only 10 percent of the new migrants to our wilderness counties were over 65 years of age. The migrants were young, highly educated professionals who generally had good incomes.

One rationale often given for why people move is dissatisfaction with where they have been living. Flight from crime, pollution, congestion, and other urban ills serves to "push" people out away from urban areas. Contrary to such expectations, we found that most people moving to wilderness areas were not overly dissatisfied with their former places of residence.

People moving toward wilderness were more heavily influenced by the attributes of the wilderness areas themselves. They put considerable importance on scenery, outdoor recreation opportunities, environmental quality, and pace of life. No single factor dominated in their decision to move. Only 27 percent gave employment as the major reason for their move. This does not fit the economic model well, suggesting as it does that income was not the driving force behind their decision to move.

Contrary to the economic theory of migration, almost 50 percent of the migrants reported lower incomes, only 28 percent had increased their income, with the rest showing no change. Recall that these are primarily younger migrants, almost all of whom had found jobs. These are not social dropouts who move into an area and put added stress on the social welfare system. Indeed, the unemployment rates in these wilderness counties is well below the national average.

The actual presence of wilderness served as a magnet that attracted people to these areas. Seventy-two percent considered it a major factor in their decision to move to the county. Among long-term residents, a majority (55 percent) also felt wilderness was an important reason for living in the area. The importance of wilderness was emphasized by the expression of the desire by a majority of both migrants and residents to have more access to these areas, and by the fact that 60 percent of the newcomers felt there was a need for even more wilderness nearby. This can be explained by the use of wilderness at least 12 times a year by more than a third of migrants and residents.

Ironically, one of the reasons for setting aside wilderness is its limited use by people and industry. Restrictions have been implemented on the amount of pollution and industrial development permitted in areas adjacent to wildlands. Given the importance of quality-of-life factors in why people move toward wilderness, there is no reason to expect such trends to diminish. Even though there was a decrease in the intensity of movement to wilderness counties during the early 1980s, partly because of the recession, the 1990s have seen a return to rapid in-migration of people into these areas. This movement into wilderness and other public lands counties has put pressure on land managers to try to balance the demands for more access to wilderness with the preservation mandated by the Wilderness Act.

The pressures placed on federal land managers in wilderness counties may be unique. Indeed, a number of persons have confronted me with statements like this: "So what did you expect? These are areas with lots of amenities. Why wouldn't people want to live there? The woods are full of loggers, hippies, code writers, agricultural herbalists, software developers, poets, gun freaks, retired folks, and all sorts of people who realize that life is more than chasing the almighty buck." My weak response usually is that what seems obvious to you has not been obvious to the academics and "experts" who educate your children, get hired as consultants, or appear as "talking heads" on talk radio or trendy news programs.

The West is more than just wilderness. My region, the Palouse, contains parts of eastern Washington, a farming area not known for its amenity values unless you happen to like wheat fields. Yet, in the 1990s, population has been increasing dramatically. Geography explains some of it—for example, people moving from the Seattle area in search of a more rural lifestyle. Despite Seattle's recent ratings as one of the most livable cities in America, local newspapers began running features on its growth, pointing out that it was becoming too congested and too attractive to runaways and the homeless, that there was an increase in crime and a general decline in civility, and repeating the usual litany of urban problems. As in other cities, living there has become more costly as the price of housing and other necessities, as well as such things as tickets to see professional sports teams, have increased, making it easier and more necessary for people to move eastward over the Cascade Mountains, down the slope, across the mighty Columbia River into eastern Washington.

Coming from similar as well as different directions, people moved into western and then northeastern Oregon, across into Boise, Idaho, another city lauded for its amenities and growth potential. Western Montana, with mountains and a winter "banana belt" for those accustomed to real cold, continued to attract newcomers, as did the towns and stark deserts in Nevada and Utah.

What most of these people moving into the interior West had in common was that they were moving within the region. Most were moving around within their state or moving to adjacent states, coming from larger cities. Geography Rules! Most of the population and large cities in the West are in California. Given the sheer number of people in that state, its problems, earthquakes and other natural and social disasters, California quite naturally has more people willing to move. Therefore, most of the migration into the interior West is from California.

The movement out of California has received the most attention and has been called a "Diaspora," a California culture in the foreign lands of the interior West. These migrants have been stereotyped as importers of Californian values into the rural West. The diffusion of the cappuccino or latex cowboy into towns of Marlboro cowboys has become a popular image in the press. "Let them drink lattés" has been the rallying cry of some as they confront the values of the old West.

The values of the old West have been stereotyped, popularized, and romanticized as well. The strong, silent, but sensitive males stand out in sharp contrast to the sometimes hard-drinking, violent, women-beating Westerners who show up in the social statistics of many towns. The high rates of violence in many Western towns often is underreported or overlooked.

Several other recent surveys in the West show an amazing similarity in why people moved there and what kinds of lifestyle tradeoffs people made. A survey of over 1,500 people in Oregon found that noneconomic reasons were the primary ones for their moving to Oregon. Only 30 percent said they moved to Oregon for employment-related reasons, about the same percentage as in the wilderness survey. Again, jobs were a minor factor in the decisions. A majority of migrants had lower incomes after their move.

Another survey of migrants in a six-state area in the Pacific Northwest and the Rocky Mountains found once again that employment opportunities provided only about a third of the reasons for why people moved to or lived in the area. People were also asked how many had moved first and then looked for a job. About 28 percent took the risk of moving to where they wanted to live and then worried about a job later, suggesting that passion of place came first. There was no careful weighing of the traditional economic costs and benefits before moving. These women and men decided that being where they wanted to be was the most important benefit they could give themselves. Regardless of whether they moved with jobs, they came for reasons related to the social and physical environment either in the form of access to family and friends, pace of life, outdoor recreation, or landscape, scenery and the environment.[11]

Academics, like other people, get set in their ways and theories, and resist change if it means acknowledging that much of what passes for "truth" is, at best, only partially true. The migration of people responding to the dictates of a "calculus of income maximization" is an oversimplification of reality. Nor should we jump to the opposite conclusion, that about 70 percent of all people who move do so for non-economic reasons. These surveys look at only a particular group of migrants, those moving to either smaller Western urban areas, cities, towns, hamlets, and rural settings. These are people moving largely by choice to what they consider to be desirable places.

There are plenty of people for whom the decision of where to move is made mainly out of necessity, where a job (and, at times, any job) predominates, and rejection of place is a luxury they cannot afford. The transformation and restructuring of a dynamic economy such as ours makes "pushed" or forced moves by blue- and white-collar workers ever more likely in the future. But for the migrants in these surveys such problems are not their main concern.

## CAN NEW MIGRANTS AND LAND MANAGERS "INTERFACE"?

Large institutions, public and private, make up their own language. Computer technology and the Internet has added much to our current language. New people moving in and around wilderness and other public lands become an "interface" issue (agency jargon). New people can create a number of different problems for land managers, especially where the national lands interface or adjoin the homes and expanding communities of these new residents.[12]

These are the types of problems that are not easy to ignore because they won't go away and they become increasingly evident as more people keep flocking in and settling on the edges of public wildlands. Land managers have to deal with more and more new neighbors who bring their own values, ideas, and expectations about what federal land managers should and should not be doing.

Tensions will heat up as private lands adjacent to public lands are subdivided into smaller multiple plots. The human use of the Western landscape historically has been dominated by small towns and their subdivisions in the form of houses and trailers, a scattering of cabins, and large farms and ranches. Unlike the spreading amoeba-like shape of a more-or-less continuous urban-suburban megalopolis sprawl, the West has a vast amount of space between places. The private space is being slowly broken down.[13]

If given a choice, people in cities prefer accessibility, being as close to work as possible and still being able to find affordable housing, good schools, hospitals, and a variety of services that meet their needs. People moving toward wildlands often display the opposite tendency. They want to be away from it all. Inaccessibility becomes a virtue.

In their previous suburban-urban environment they may have lived close to the on-ramp of a highway. Now there is no on-ramp and they are glad to live on a remote, poorly maintained dirt road. The lure of wilderness leads to a desire to live remote from other people. Where as they may have previously lived cheek by jowl with their neighbors, now they prefer to have as few neighbors as possible within "spitting distance." The closer to wilderness the better, since it provides amenities and guarantees few neighbors. The big neighbor is the federal government, and its actions can create a wide range of conflicts.

Fire protection is one of the more visible conflicts for which hundreds of millions of dollars can be spent. Nowadays, fires in wilderness areas are considered part of the natural process and are normally allowed to burn themselves out. But suppose you buy a house outside a wilderness area, and a fire starts and threatens to destroy your house. The natural tendency is to shout and demand protection of your property. Is it the agency's responsibility to put out these natural fires in wild areas? Most would probably say to you, "No. You took the risk, now you take the consequences."

The shouting for protection gets a lot louder when the houses—or in many cases weekend or summer cabins—are located on the edge of a multiple-use production forest. Not only do the people who bought or built residences outside the forests demand that they be protected, so do the potential buyers of the standing timber who don't want to see it go up in smoke. The news clips show massive fires with shots of smoke and flames threatening houses and towns. People are flown in from all around the country and paid high wages to fight these fires.

The scenes of women and men fighting fires convey visions of bravery, solidarity, and the nobleness of trying to combat the destructive forces of nature. Too often the questions are ignored of whether fighting these fires is worth the risk of human life, what the value of the houses is, and whether the action is in accord with the "let burn" philosophy and science of some agency scientists and managers. For example, during the summer of 1994, the cost of fighting fires was over $30 million for Idaho alone, nowhere near the value of the barns, cabins, and houses that were threatened by these fires. It seems pointless to spend $30 million and risk lives for cabins and houses that are worth far less than $30 million from fires.

In Colorado, where the risks and costs are high also, individual owners who moved into high fire-risk areas now demand protection or compensation. Do they have that right? The tendency to fight fires if they threaten property has another perverse effect. It is an open secret within the agencies that some fires are deliberately set by persons who count on fighting summer fires to earn part of their income.

The fire issue is one where newcomers and old-timers alike feel that the government they love to hate has to protect them from the hazards of nature. It is their right to demand such protection. Why it is their right is not clear. The agencies' primary role is to manage land, not to serve as a local fire department, though shaking Smokey's admonitions about putting out fires from the American psyche is difficult to achieve. Not putting out fires goes against the grain and creates guilt much like not reporting a fire and letting a house burn.

If people living by wildlands demand protection from natural hazards, there are any number of actions that they don't want land managers to continue. Timber harvesting, particularly in the form of large unsightly clear-cuts, is unpopular with many new people. They did not move to their new home to see huge swaths of forest cut. The serenity of hikes through the woods is not improved by viewing the results of industrial forestry.

Tensions mount between newcomers and the old-timers if timber harvesting is symbolic of a community's identity; for instance, the local high school team may even be known as the "loggers." For old-timers, clear-cuts and logging may remain important symbols of paychecks and jobs, just as smokestacks of Eastern cities in the 1950s were promoted by local chambers of commerce as vital signs of prosperity.

The slogan "Idaho Is What America Was" that is seen on t-shirts has more than one set of meanings.[14] The illusion of a safer, kinder, family-oriented America set within a beautiful and bountiful nature needs to be balanced with the images of private timber companies overcutting private lands and, because of their own wastefulness, becoming dependent on public lands for supply. Until recently many federal land managers have seen it as their duty to supply the raw materials of minerals, timber, and land to private individuals and companies.

The new migrants may move toward the wild so that they can enjoy the laid-back lifestyle hinted at in the "What America Was" slogan, but they don't want to live with the consequences of industrial management of the public lands. The simple day-to-day experience of driving down one-lane roads with logging trucks hurtling down at them has struck fear into many newer residents. The uses of logging roads increase as people move into forests and want services such as garbage collection and other prosaic utilities common to everyday life.

The newcomers may oppose a number of other "normal" policies of the land managers. The spraying of pesticides and herbicides often is cited as another "interface" problem. People living near the forests simply don't like or fear the consequences of planes spraying toxic chemicals, with the attendant dangers of having those chemicals drift onto their property. Environmentalists or not, they often simply don't believe the claims of land managers that such actions are both safe and necessary to protect trees from infestations of beetles, fungus, or whatever parasite might be posing a threat to trees.

While they don't like current management practices, newcomers may put new demands on the wildlands by treating them as their personal recreating and playground areas. They may not be satisfied simply with using the available hiking trails, but may demand more and better-maintained ones. For example, there has been a dramatic increase in the use of mountain bikes. Restrictions on their use in wilderness areas can lead to complaints from bikers that they are being treated unfairly and should be given more access. Not only the availability of access, but the types of activities allowed, and where and how, can lead to problems for groups with different interests.

Hunting is an accepted practice on public lands, except in national parks. It is also a way of life in many communities throughout the West, for both men and women. For some, it is the main source of meat on the table throughout the winter months.[15] Hunters, whether newcomers or old-timers, expect to have access to these lands. For long-time residents, hunting season is like the seasons of the year, inevitable and predictable. New migrants living in isolated houses or within new developments can be traumatized by shots ringing out nearby or on their land. Demands for limits on how close people can hunt on lands near homes are not met with favor.

Hunting accidents are treated like auto deaths; they are unfortunate, but inevitable. In the West the chances of going to jail are greater for a drunk driver who commits vehicular homicide than for a sober or inebriated hunter who kills someone.[16]

Hunting conflicts are not the only kinds of neighbor-to-neighbor conflicts over the use of public lands. Recreational uses may be incompatible, or different users may demand more of a particular use of the public forests. Cross-country skiers do not want motorized ski machines blaring at them or disturbing their tranquility. They say they were there first. Off-road vehicles have been gaining in popularity throughout the West. If they outnumber cross-country skiers, should they be given increased priority? Should they be allowed in non-wilderness areas being managed for wildlife or watershed protection?

Non-motorized uses can create problems as well. New resident hikers may not want to share trails in wilderness with guides and outfitters who use horses and mules to carry people and supplies to camps. They may not want to run the risk of stepping in horse/mule manure, or even more trendy llama "droppings." Land managers have to consider how, when, why, and where to separate and segregate different users of these lands.

People living on the edge of wildlands also create problems for managers by how they use their own private lands. Hunting is but one example. It can be troublesome when "new" people deny traditional access to hunt on their land. Their private lands may also have been used for access to the public lands. They may bar access, making it necessary for the courts to determine whether such actions are legal.

Newcomers moving to the wide-open spaces may dislike the open range aspects of the West. Finding your neighbors' cows in your yard or garden leads to demands that the animals be fenced in. Never mind that the cows were there first. The more newcomers within cow distance of the grazers, the more solidarity builds for demanding changes, even though most western states are open range states. If you want to keep the cows out, you have to fence them out—not their owner. Land managers who supervise the leasing of public lands for grazing get caught in the crossfire.

How people use their private lands may impact the watersheds and ecosystems land managers are charged with protecting. Increasing numbers of cabins or houses in developments with septic tanks can pollute streams, endangering fish and wildlife and human water supplies. Newcomers who demand fire protection can increase the threat of forest fires from burning trash, if the fire gets out of control and spreads to public lands.

The demands on federal managers to "help" their neighbors is controversial. If they are supposed to serve the broad public interests, to what extent should they cater to the demands of an on- the-spot vocal minority who in any event are getting a lot of free benefits from living near public lands. Yet they feel they have a greater claim simply by virtue of living where they do.

The "interface problem" as defined by land managers exists in large part because of the movement of people into the area. And not just the number of people, but the type of people. Previously land managers had fewer neighbors to deal with, and they often had similar objectives: help with the harvesting of timber on these lands. The mindsets were very similar, and townsfolk and government managers got along well.

The newcomers create a more diverse group that wants the land managers to manage differently. They want to live near a protected landscape, not the old commodity landscape. The newcomers not only fragment the landscape, but the local culture as well. Still the change is only beginning. Even while bringing different values, almost all of the newcomers are white, and generally well off or have chosen a genteel, lower-income lifestyle.

Much of the West from where they will increasingly come is becoming more Hispanic, Asian, and Black. While it is presumed that all groups in multicultural America share common values, there are also differences rooted in cultural heritages. How these other groups would want to use the wildlands of the West is only beginning to be considered by land managers. Current conflicts and the inability to resolve them do not bode well for the future.

The evidence of ongoing conflicts between land managers and new and old residents of surrounding areas indicates that management has not kept pace with social, cultural, and economic change throughout America. A continued inability to act aggressively could result in even greater costs in resources, lost opportunities, and agency credibility. The very economy of the West within which they operate has been changing. Until recently they have been unaware, unsure, or unable to respond to these dramatic changes.

# Chapter Eight

# Wilderness and Economies of the Old and New West

There is what is by now almost a cliche in economics that says, "Practical men, who believe themselves to be quite exempt from any intellectual influences are usually the slaves of some defunct economist". Madmen in authority, who hear voices in the air, are distilling their frenzy from some academic scribbler of a few years back." Certainly, whether anyone is aware of it or not, almost all economic policy in the U.S. is driven by the ideas of John Maynard Keynes, who made that famous statement. The Keynesian Revolution has been at the heart of European and American economic policy decisions. Much the same can be said about regional development policy in what to most Americans was an obscure debate between two economists in the 1950s.[1]

It is not so much the debate itself, but the view of what drives the development of regions that is especially relevant to the economic future of the West. Douglas North, an economist at the University of Washington, argued that regional economies were driven by their export industries. He used the historical development of the Northwest as an example. The Northwest by definition at the time extended into today's Rocky Mountain region. With the coming of white settlers into the region, wheat, flour, and lumber were quickly developed as exportable commodities. The initial markets were in California in the 1840s. The demand for wheat and lumber increased tremendously with the gold rush in the West. By the late 1870s, wheat from the Pacific Northwest had become a vital part of the world wheat trade. Ships carried the region's wheat around the world to England, Europe, Canada, Japan, and Australia.

The growth of the lumber industry followed a similar pattern. The first shipment in 1847 went to California, with exports expanding rapidly during the gold rush. Markets expanded into the American Midwest and continued to do so at the expense of the Southern timber industry. By 1900 lumber and flour accounted for almost 60 percent of the value of the region's manufacturing output. Other areas in the West followed suit, exporting products such as furs and minerals around the world.

What drives the development of a region is the demand for products that it exports to other regions and the world. The regional economy can be divided into two major groupings, the export industries and the "residentiary" industries (those that produce for the local market and the people in the area). It is the export industrial base that is vital in determining the growth and income levels in a region. The "residentiary" or nonexport base part of a regional economy is passive and dependent on the growth of the export-oriented industrial base part. For the West, as goes the demand for its resources (lumber, wheat, lead, silver, etc.), so goes its economy.

Given that the growth of a region is almost totally dependent on the success of its export base, decline and change in the export base must be met by the growth of other exportable commodities or goods, or a region will be left "stranded."[2] This decline has to be counteracted by finding or developing new export-based industries. If not, dire consequences will befall the area.

The West is littered with old ghost towns that were left behind after the gold rush, or more recently with energy towns that were built on extracting coal or oil. When the demand for their export products disappeared or the resources became exhausted, or other disasters struck, the towns went into decline and never recovered. The boom-or-bust cycle has been part and parcel of the history of many places in the West.

Economist Charles Tiebout responded to North's article by arguing that there was no reason to assume that exports are the sole or even the most important factor determining regional growth and income. The nonexport or residentiary (local) industries would be a key factor in any potential development of a region. There followed a gentlemanly exchange during which neither essentially gave way on their position.

I have introduced this academic debate between two respected scholars not to get into the fine points of their respective views, but rather because the ongoing debates over timber, salmon, and other resources have these debates lurking in the background. In a sense, North won the debate because his view prevailed. In the West, the importance and logic of the export base theory has been accepted and internalized in public lands management. Decisions have been, and continue to be, made assuming that the local and regional economies depend on having a sustainable export base. The view has become so pervasive that it is accepted by people on various sides of the development debate, from free market theologians to market-constraining environmentalists.

At the local community or county commissioners level, the export-based approach has been accepted even more heartily because it seems to reflect what they and their "pioneer" ancestors learned from their own experience. When trees were cut, wheat was exported, and lead, silver or gold were extracted, times were good. When they were not, times were bad. Local development is based on keeping good times going.

The computerized models of today only verify what is obvious. When exports decrease, local economies go down. Indeed, the models are extensions of the export base, and so the results should not be surprising. Unfortunately, the export base model reflects not only an older West, but an older America as well. The local citizens and politicians, as well as students and scholars, would have been more farsighted if they had listened more closely to Tiebout than to North.

## MANUFACTURING AS THE PREFERRED EXPORT BASE

When I first taught about regional economies of the United States at the University of Texas, export base theory appeared logical. The

University of Texas at Austin, the second most wealthy university in the country,[3] collected royalties from state lands that sat on oil deposits. It was the basic assumption of many high rollers in the state that the economy had and always would run on oil. Students, most of whom came from Houston (the real symbol of oil), Dallas, San Antonio, and other urban areas, understood the economic role of oil better than I did. The rest of the West was more confusing to them.

When I described the economies of the rest of the West, some but not all of my examples were obvious to my students. They understood the classic examples of ranching. After all, the Texas mascot was a longhorn steer. However, their knowledge of forestry was more remote, given the absence of forests and trees in central and west Texas. The descriptions of logging industries in the Northwest had to do with far off "other places." Another industry that formed the backbone of the Northwest was aluminum production, which depended on cheap and subsidized electricity. The need for water was fairly obvious, given the arid nature of Texas. Mining was more exotic, but the quest for riches hit a nerve with most students.

Having discovered and extracted resources, the next step is to process them further. The need to add value to raw materials in the West via manufacturing seemed obvious. Simply extracting raw materials is what Third World nations do. An exploitive relationship develops between colonizer and colonized. Indeed, historians of the American West have used the same arguments in many of their writings. Only it wasn't the foreign European powers exploiting the African, Asian, or Latin American countries. It was the Easterner exploiting and extracting the physical and human resources of the West.[4] The exploitation of resources led to using those resources to manufacture products for use elsewhere, having exported them out of the region. Manufacturing would provide the future engine of the Western economies. After all, manufacturing and exporting were the mainstays of the local, regional, and national economy. The West needed to catch up to the rest of the country. Whether in an urban or in a rural location, the approach to development required the promotion of export-based industrial economies. Manufacturing produced higher-priced products for export.

The image of the success of local economies in the West is often the "sweet" smell of pulp plants processing raw logs into consumer products. The next step after cutting trees or hauling minerals out of the ground is processing them in industrial complexes in rural areas. It is these industries that contribute to the growth and development of the region. There is a reason why smokestacks are still welcome in these areas.

The other side of export base theory is that industries and places are not static. The larger cities in the region cannot succeed by remaining tied to wheat, forestry, mining, or even airplanes. Either their economies change, largely by becoming more diversified, or they wither over time. When they do so, these older industries do not simply go away, they move out of urban areas into more rural places where the cost of labor is lower. They follow a product life cycle.

The rural places, in the West or elsewhere, get companies when they are on the downcycle of their product life. When they have become routinized, less innovative, less entrepreneurial, no longer needing the dynamism and brain power of urban areas, they are ready to move out to the provinces. The newer, innovative industries locate or stay in urban areas. The trickle-down effect assures that only standardized products requiring less skilled labor will be manufactured in rural areas.

Prospects don't look good for the rural areas of the West, since by many accounts they are resource driven and exploited places. Added to that, they don't at first glance have many entrepreneurs, the new "folk heros" of American industry. The entrepreneur, not the rancher, logger, or cowboy, is considered the mainspring of economic development.[5] Urban areas have the concentration of innovators, capital, and information on which new firms thrive.[6] This implies that, despite recent migration trends and advances in technology, rural areas will remain behind urban areas in development. The reliance on extractive and related management industries means that highly educated people with special skills will not be required. The type of development will not be of the right kind.

## TOURISM: SELLING IMAGES OF THE WEST

The future image of the West, particularly in the high amenity places, is of dependence on another type of export industry, tourism and recreation, a playground-oriented development strategy. Bring us your tired and worn out workers, rich or not-so-rich, and we will provide rest, relaxation, fun, excitement, and physical challenges. Tourism has always been a part of the economy of the West, and it is becoming a more important part, but can it shoulder the burden of the New West?

Tourism is not an attractive development strategy to loggers and cowboys. It implies servility, low pay with little machismo, and moving from working in the outdoors to indoors. Having to flip hamburgers at McDonald's or some such place is among the biggest fears of what Westerners might be forced to do. Tourism also is stigmatized because it caters to outsiders, creates economies based on either myths and images of a frontier West or on recreation, not on the reality of providing what people really need: wood products, potatoes, and gold for rings and chains. People in the West consider the work they do to be rooted in dirt and water, not images promoted to attract tourists.

Tourism, with its motels and promotion of false "friendliness," modifies and often creates a very different atmosphere from the one that used to exist. I often feel a sense of sadness when I visit these tourist-dependent towns. The older authentic towns have been transformed with fake storefronts, fake saloons and jails, and stores selling Western clothes that no Westerner would wear. They cater to an image of what the old West was perceived to be, not what the "real" town was like. Or they sell themselves as Swiss or Bavarian mountain villages. The tourists pass through, looking, buying, almost in an assembly-line fashion. No wonder no one wants to be considered a tourist.

Focusing on tourism as a development strategy creates an unreal world. The lakes and mountains of the wilderness serve as a backdrop, not as an integral part of a travel experience. Increasingly, the distinction between tourism and shopping becomes less and less clear as shops and mini-malls compete for more and more of the tourist's time. As a development strategy tourism also suffers from the need to market uniqueness. Whether built around images or intensive recreational uses such as skiing, there are only so many tourist towns that can successfully create a niche and market for themselves. Ironically, while many Western towns have a unique character, efforts to make them more "attractive" to tourists often make them more alike, and they become places to visit and to spend money rather than to live.

## THE ROLE OF SERVICES, OR WASHING AND EXPORTING YOUR LAUNDRY

The most dramatic change, both at the national and regional level, has been the rise of the service sector of the economy. The growth in jobs nationally has been in the services sector due to both increased worker productivity from restructuring of manufacturing in the U.S. economy, and continued movement of manufacturing jobs outside the United States. Recall from the North-Tiebout debate that the service or residentiary sector was considered to be reliant on the export or manufacturing sector. It is estimated that each manufacturing job "creates" an additional two to four service sector jobs.

If a manufacturing plant comes to town and creates 2,000 jobs, then it results in a demand for a host of other products in the service sector such as haircuts, food, automobiles, or laundries. If the manufacturing plant did not come to town, there would be no need for those additional services, and so the services sector wouldn't expand. You can't create an economy simply by people taking in each other's laundry, or buying various services from each other. At least that is how the "manufacturing as the engine of development" theory goes.

This argument began to appear very suspect as the service sector grew, while at the same time the manufacturing sector contracted or grew at a slower rate. Using the manufacturing logic, as the manufacturing sector declined it would support more and more service jobs. This does not make much sense. For example, in the wilderness counties the total number of jobs was increasing and manufacturing jobs decreasing over the last 20 years. The fastest growing sector was services. Throughout the interior West, as the resource industries suffered as the result of job losses, the service sector was expanding and supporting a larger and larger segment of the region's economy.

Why the service sector has grown so dramatically is not so obvious. Throughout the West people refuse to let go of the belief that their economy is driven by the timber mill, the rancher's cows, or tourists. Actually they provide only a small part of the local and regional economies.[7] Yet with a few exceptions, people's perceptions do not match objective economic reality. When asked to describe the importance of various sectors of the economy, people in rural areas will cite the resource industries such as agriculture and forestry as being very important to the local economy, while in reality they account for a minor part of the jobs and incomes in their counties.[8] The growth in jobs and local income is driven by services and other related sectors.

A major reason for denying the importance of the services sector is the perception that flipping hamburgers, pumping gas, or cleaning motel rooms are the major types of jobs in this sector. But, in reality, the services sector is very broad and includes a wide variety of jobs connected with the Internet economy. People who write code or work as software engineers are part of the service sector, as are accountants, lawyers, and other jobs connected with the information age. These types of service jobs are not related to the growth of manufacturing; they create growth rather than being derivative of it. They are highly skilled, highly paid jobs, not the traditional low skilled jobs insinuated by the term "service industry."

The image of low-paid service jobs is not accurate for much of the West. People moving and working in the wilderness West often are well-paid professionals, quite different from the laid-off timber workers.

Another misconception of the service sector is that it is not export oriented: products are all locally consumed. Clearly in this age of linking with and sending material via fax, computer modems, and the Internet, as well as by overnight delivery, work in the service sector is not limited to specific places. Accountants, lawyers, computer and software specialists, and others can sell services worldwide both from large urban areas or small towns next to wildlands. Recent studies show that about one-third of producer services are exported out of the area. This figure seems to hold whether you are in Seattle or in more remote wilderness counties.[9]

The growth of the service sector has been taking place both in urban and rural areas, and there is evidence of a movement of producer services out into the rural areas. For example, from 1974 to 1986, service employment growth was the dominant source of employment growth in the rural and wildlands of the Northwest. Information services is one of the key sectors growing in the rural West. The service sector can consist of many of the types of high-technology jobs normally associated with high-tech manufacturers.

## CHOOSING THE KIND OF GROWTH TO PROMOTE

With the urging of organizations such as the local Chamber of Commerce and local economic development councils, the towns and small cities of the West are increasingly encouraged to do some kind of promotional activities to foster economic development. Even the smallest villages engage in some form of growth promotion. The efforts intensify when timber mills shut down, and are encouraged by agencies such as the Forest Service, which considers that part of its mandate is to help keep communities stable or to help them change.

Certainly it is important for citizens of towns and cities—no matter how small—to consider what they want their community to be like in the future, what they like and want to maintain, how they want to change or contend with economic realities. Controls and planning are not popular in the rural West. But without them, citizens are even more impotent in having any say about the future of their community.

The reality is that, whether assisted by the Forest Service or not, communities are unable to have much influence on the national and international forces that shape their local economies. Settling up local development groups has not been an effective means of generating local jobs. Nor has the "giving away the store" approach of granting local tax breaks, providing free land, setting up industrial parks, and restricting the power of unions been a sure way to attract new companies to a community.

Nonetheless, the common approach continues to be to find a manufacturer that is a branch plant of some large company—preferably a high-tech company—to lure to your region. Chasing after companies is a practice that all government entities, from small towns to states, are engaged in. Most states have a department of commerce whose main function is to attract industry. Rural areas trying to attract business and industry are essentially participating in a lottery with very long odds. The number of actual plants relocating is tiny compared to the tens of thousands of local governments trying to win the growth sweepstakes.

As in the case of a Northwest computer company that recently expanded, the sweepstakes are televised when a company announces that it wants to expand and build another plant somewhere, bringing in several thousand jobs, and will accept proposals from communities and states. The odds of winning are long, and if you win you may find out that the costs of the tax breaks and other sweeteners, when added to costs of supplying the basic services such as roads and schools, are not offset by whatever financial benefits the company may bring to the community. In rural areas, the scale of development can easily change the character and the perceived quality of life in a community and region.

The evidence from large pulp mills near public lands shows clearly how they can contribute to the instability of an area. These large companies are also at the mercy of economic shocks and random events. Declines in the price of timber products leads to layoffs with little notice. Even if prices are high, competition forces an increase in labor productivity via mechanization and a decrease in the number of people employed.

Blame may be placed on a lack of supply from national forests or on environmentalists, but mostly this is just scapegoating. Forces outside the control of local communities or federal land managers, such as changes in interest rates, deficits, and tax structures, can wreak havoc on these rural areas, especially if the companies they give their allegiance to are not dynamic, competitive, or capable of adapting to change. Unfortunately for the rural wildland communities, timber companies often are marginal firms that, without subsidies or high prices for their products, wither under the threat of competition.

Historically, resource companies have not provided a stable base for the West. To an outsider, the people of the West seem to be putting their faith in the industries that have let them down time after time. Why? Are they another example of the old adage about people who do not learn from history? Why continue to put your faith in a flawed or failed strategy? Why continually trust those who provide an unstable job environment? The answer, beyond the rhetoric of Us versus Them—loggers vs. environmentalists, locals vs. outsiders, West vs. East—is that the people who live there do not do so simply to harvest trees, raise cows, or go a mile underground to search for silver. They hold onto the old economies because of their dedication to a way of life. They share several things with the people who are new to the region, many of whom share no affinity with the economies of the past. They want to be able to live where they choose to live.

## REGIONAL DEVELOPMENT AND A GEOGRAPHY OF PLACE

Almost all of the research, theories, proclamations, and politics associated with promoting local or regional development is obsessed with the economic dimension of our lives. Some of the theories are obviously too simplistic, focusing on one or two basic factors such as exports, and assuming that they can explain what is happening as communities and regions experience either growth, decline, or stagnation. Clearly, regional economies are much more complex than the way they are described by old theories made new by running them on computers.

That local growth is necessary for prosperity is an assumption that goes unquestioned in too many places. That is also true of the notion that because a particular economy has provided a community identity, it must continue to do so in the future. Particular industries such as forestry get associated with providing a cultural basis to Western economies.

This is not so. It assumes that people in the community live there because they want to cut down trees. Evidence for this might be the migration of loggers to where the trees are. There has been increasing competition from the South in providing timber. All indications are that, because of climate and the ability to grow trees faster, the expansion of the American timber industry will take place predominantly in the South.

Most loggers in the West just won't move to where the jobs are. They want to stay in the communities where they have made their homes, surrounded by the forests of the West. They live where they do because they want to be there and want to live in a particular way. Their cultural identity is tied in with the landscape of the West, and how they use their time working, playing, and living in these places. The stereotypical view of the life of a logger or other Westerners is of a life revolving around driving pickups, hunting, fishing, or just plain roaming around the public wildlands.

Most theories of regional development assume that people seek higher wages but, as shown earlier, most people in the rural West do not. Or they feel that an amenity-rich life built around a landscape compensates for the lower wages. Instead of focusing on wages as the main goal in life, it makes more sense to realize that wages are what people trade off for a particular lifestyle that they think comes with the territory. When asked if they were less stressed and happier when they moved to a wilderness county, even though they received lower wages, the vast majority said they were. People who move toward wilderness and stay once they get there feel very content and satisfied with their decision to do so.[10]

One reason economic models don't work well in explaining migration is because they don't consider the substance and context, or how people want to live their lives. The real income people have after a move may be roughly the same as before the move if quality of life is taken into account.[11]

Any theory of local or regional development should consider the attachment people have or develop for particular places or the elusive sense of place that many writers have been trying to describe. It is the sense of place that keeps people loyal to rural landscapes which in return rarely provide them with much of material value. It is commitment to place that keeps people from moving even when the mines shut down, the timber companies cut back or close down, or ranching provides only a small yearly income.[12] Economies and cultures evolve together. Economics assumes no commitment and culture assumes an attachment that goes beyond economics, yet they are often linked together.

How does a local or regional identity develop? What is special about a place that, despite economic hardships or incentives to move, makes people unwilling to leave? Often what makes a place unique is not provided by the market, and so it is usually ignored. Uniqueness in the West comes from a clean environment, lots of public open space, wilderness, and friendly neighbors. These non-market "qualities" often are compromised and destroyed as development takes place, leading to a loss of uniqueness of a particular place.

To maintain the uniqueness of an area such as the wilderness West, in a dynamic sense, is to increase its viability by development that maintains or improves the quality of life by fitting harmoniously into the natural and cultural environment. This is easier said than done, but almost all models of regional development ignore such issues. There may be a lot of emphasis on sustainable development, but few practical applications.

The emphasis on the uniqueness of communities in the West comes primarily from its residents and their sense of place. The area's vitality depends on having a sense of place. It recognizes that the need to be rooted is one of our most important and least recognized needs.[13] How do you get roots? By virtue of your real, active and natural participation in the life of a community. In part such participation is automatically brought about by place and social surroundings.

Areas such as the rural wilderness can create a more developed sense of place because of the closeness and interplay with nature.[14] A stronger sense of place means a stronger community, which implies a stronger local economy. A community, by definition, has a particular location, and its success cannot be separated from the success of its place, its natural settings, and surroundings.

The two economies, the natural and the human, support each other and the hope of a continuing life. In both agricultural and timber-dependent areas, development or "progress" often is possible only at great cost to the environment. This progress tears at, and ultimately may destroy, the fabric of place. This gives some insight into why, even though we live in an urban society, we still say we prefer small towns. Urban society for many has broken down. Moving toward and living in "wilderness" provides us with the hope of recovering our roots and a sense of place.

A sense of place comes from, or is derived from, local cultures, and not from the economic structure of a place. This raises the question: "Don't jobs and a local economy come first and then a local culture emerges?" Did not mining and lumber activities get people to move West? The people stayed and, because of these extractive industry jobs, a specific sense of place developed through particular community personalities and social structures. Indeed this may have been the initial process as non-native, mainly European peoples settled and spread over North America.

The last thirty years make such a scenario more and more unlikely. The job basis for these communities has changed, with jobs more likely to follow people, and people more likely to move to or stay in places where a sense of place is rooted in the social and physical environment. It is the sense of community and lifestyle, and not jobs that give meaning to their lives. Or at the very least, the setting within which people do work at their jobs becomes increasingly important. This applies both to the loggers and miners who don't want to leave the woods and be forced to do low-skilled jobs, and highly educated "infotech" persons who want to live and hike in the same woods.

The sense of place developed from interacting and living with nature, which is hard to do in the primary landscape of most Americans—the suburb. That is why there is such prejudice against the suburb (by academics and planners, anyway). To the casual eye it is a rootless landscape, unanchored, and hard to differentiate. Many of the mass-produced suburbs throughout the nation create a homogeneous landscape and a loss of character.[15] Consider Deborah Tall's poignant observations:

> Like so many Americans raised in suburbia, I have never really belonged to an American landscape. The narrow strips of spared trees buffering my several childhood housing tracts from nearby highways don't qualify as much of a landscape. Nor does landscaping, clumped shrubbery from the nursery transplanted under maternal directives on Saturday mornings—a row of squat evergreens screening the house's cement foundation. Bulldozed, paved, it was a terrain as homogeneous as the developer's desktop model. As Gertrude Stein says, "When you get there, there is no there, there." The land's dull tidiness was hard to escape, except in the brief adventures of childhood when I could crawl beneath a bush or clothe myself in a willow tree.[16]

The physical variations between suburbs are largely due to climate. The features of the natural environment fade into the background. There is a loss of sense of place, at least as it is derived from the physical environment. The scale at which one can interact with nature is diminished greatly, if not lost completely.

Much of what makes the West unique and gives it a sense of place is the physical environment and its elements of the wild. It is the sense of place that is important to the social and economic well-being of people and their places, not whether they continue to work in extractive industries. Indeed, it is these very industries, in their "using up" of the landscape, that contribute to the destruction of the Western sense of place.

The historian's depiction of the West as a colony is based on the role of outside money, the railroads, and corporations that came West but did not remain. Extractive industries such as mining and lumber are subject to classic boom-and-bust cycles in any event. They have only limited control in terms of competition from other regions and countries, as well as the rise and fall of prices for these commodities. Extractive industries are not the preferred way to build a vital economy that can adapt to change.

The beginnings of the industrial revolution in the 1700s in England and continental Europe were followed by a search for colonies from which various commodities could be extracted. Even after the end of the colonial period, a dependence on extractive commodities has not served many of these former colonies well.[17] It is dependence on the natural resources of the West, whether gold, silver, coal, oil, uranium, trees, or water, that has made negative comparisons to earlier colonies strike home.

Today the image of the West as a region controlled by outside private forces rings less true because of the economic growth and rise to political power of its own large cities. Many of the corporate centers of resource companies are now located in the region, but in the larger cities and not in the small communities. Others remain outside the region or outside the country. Whether a natural resources company is in San Francisco, Atlanta, New York, or Boise may not matter. They are removed from the communities they operate in, and create both dependent and fragile local economies.

Large corporations headquartered elsewhere do not develop a local or regional sense of place. Although corporations legally are considered to have the same rights as individuals, in general they do not subscribe to the same moral code as individuals nor are they held to it.[18] Corporations do not benefit from wilderness, nor is it important to them. Imagine talking about the spiritual value of wilderness to corporations. A corporation's executives and employees might benefit from experiencing wilderness either literally or symbolically, but not the corporation itself.

The people working in the communities for large resource companies do not expect them to have any particular concerns or attachments to the place. When asked what they think about a logging or mining company pulling out, the usual response is "Well that's business, and they have to make a profit." Curiously, but not surprisingly, the same attitude does not hold when the government says it can no longer sell timber.

The search for profits is counter to the search for a sense of place because it destroys human roots by turning the desire for gain into the primary motive.[19] The most obvious examples of outside companies devastating entire landscapes are the strip mining of coal in Appalachia, and more recently the logging of timber on private lands in western Montana. There are actually whole mountains that have been stripped bare of all their timber.[20] The push to maximize profits as well as the threat of corporate takeovers resulted in some companies cutting almost all the standing timber and then selling the land. The timber represents capital; cut the capital, and the company becomes less of a takeover threat. Invest the profits from the cut timber in another business that promises higher returns.[21] Retaining their landholdings and providing for future wilderness is not among the objectives of these companies.

The cut-and-run image of some companies has created the need for others to run ads and public relations campaigns in which they present themselves as practicing environmentally sustainable logging that will continue for generations. And indeed there are lumber companies that have a commitment to their region, and to showing that logging and protection of wildlife are both part of their corporate policy.

If the history of the West does not give much reason to believe in the commitment to places by large corporations, anecdotal evidence suggests that local firms are more likely to care about local people. Local businesses will be more likely to use local suppliers, and local interconnections will be greater. If the owners and the employees are from the same community and see each other in a variety of settings, they develop a greater sense of responsibility for each other. It is much harder to fire someone you see in the local tavern or church. By contrast, it is much easier for the executive who is sitting in a chair hundreds or even thousands of miles away to fire employees he has never seen.

A place-oriented approach to regional development emphasizes a shift toward a locally controlled development. It suggests that the businesses are there because their owners, like other people, live in an area because they want to, and not simply because that is where they can make a profit. There is evidence to suggest that this is true in the West. For example, in Montana quality-of-life factors are very important in explaining why a range of businesses and industries locate there. Among the important reasons they give for choosing Montana are the "rural nature of Montana," recreation, proximity to wildlands, and quality of the environment. Similar results hold on a national level for a wide range of businesses that are located in and have been moving to rural America for quality-of-life reasons.[22]

Keeping a high-quality environment becomes important not only to people but to business. It becomes a development strategy, and has been labeled the environmental model. It emphasizes the role of the preferences of people and businesses, and puts quality of life and environmental quality at center stage, instead of off stage or in a peripheral and minor supporting role. It builds on the work of scholars and others who have argued that "geography matters," or more recently, as the popular press has pointed out, that "places matter." The old academic geographers who were not much listened to, were ignored, and even ridiculed must be celebrating. The importance of places, and what makes them both unique and desirable is where the theoretical emphasis should be, and not on the sterile economic calculus of maximization.

The realization that places and their social and physical environments are important has infiltrated the economics profession as well. Indeed, one of the places where the field of environmental economics developed was at the bastion of free market economics, the University of Chicago. Starting in the 1960s, George S. Tolley and a cadre of his students, post-docs, and visiting scholars began asking questions about the value of the environment, and how to value it.

Some of the early estimates for the U.S. Environmental Protection Agency of the benefits of cleaning up the air, water, and solid and hazardous wastes were provided by studies done at Chicago. Environmental economics became one of the newest sub-branches of economics. What had been a minor topic of the economics curriculum, the advantages or disadvantages of polluting the environment, became a major field of study. At about the same time at a number of other universities, environmental economics became part of the curriculum. [23]

I would argue as a geographer that there is a crucial need to identify with a place and its people. Without this, many of us would go crazy, or at least begin to disassociate ourselves from the larger society, whether by not caring, not voting, or practicing what normal society classifies as "deviant" behavior.[24] For an extreme example, ask any gang member in a big city if he feels he is part of American society. He will tell you in colorful terms that he is not. By contrast, ask the same question in the small towns near wilderness areas, and listen while they tell you that they are America and they love their country; it is the federal government they can't stand. And they know their surrounding physical environment intimately. I indulge myself by using extreme examples, but they are telling.

A few innovative economists have joined geographers, humanists, planners, journalists, philosophers, and even lawyers in arguing for the importance of sense of place. If maintaining a local or regional sense of place is important, should it be supported by society? This is a controversial question. If the culture of the West is rooted in timber towns and ranching communities, should they be "preserved" if they are threatened with economic change or extinction?[25]

For economists, it becomes a question of the willingness to pay for something as intangible as the value of established communities. A natural approach is to use contingent valuation, which economists have used to estimate the value of landscapes such as parks, wilderness, endangered species, and other goods not priced in a private market setting. Contingent evaluation also includes, in addition to the value to people currently living in an area, the optional value to those who might want to preserve a place because they want to have the option of moving there later, as well as the value of preserving the sense of place of an area even if they see no possibility of moving there.

Research has shown that people will pay for option/existence values to preserve wilderness and other public lands directly or indirectly, given the ongoing conflicts in the West over these very issues. The idea of extending this to communities closely integrated with wild public lands and their ecosystems should not seem far-fetched. Certainly we would not do this for all communities. It would be hard to imagine considering paying to preserve a typical suburban landscape.

The idea of using a willingness to pay to maintain a sense of place via buffer zones in and around wilderness lands may be useful when evaluating various strategies for maintaining and developing the vitality of local communities, particularly ones likely to be under stress and undergoing economic and social changes. Nonetheless, economic approaches to a sense of place evaluation will provide only conservative estimates of an attachment to place.

I have made it clear that I do not believe that the economic dimension is the most important in a ranking of values underlying a sense of place. But to not even consider as a lower boundary the economic value people will try to place on their attachments to place in the West is to ignore the importance of the relationship they have to their local culture and the physical environment within which they live.

The economic dimension of sense of place is "revolutionary" in that it recognizes the need to put developmental theories in a broader place-centered perspective. In the heat of the debate between the preservationists, conservationists, developers, and others in Western communities, it is often forgotten that the surrounding wildlands have provided a means of community identity for all citizens, both directly to those who live nearby, and to those far away. That is why the locals don't want to move away, and why most Americans do not want a privatized wilderness. Selling wilderness implies selling identity.

The importance of sense of place and wilderness is that most Americans seem to agree that wilderness and public lands are important. Even if sense of place is hard to measure or, perhaps should not be measured, most will agree that it needs to be considered in discussing how we want our communities to develop. The presence of wilderness in the West adds something both to the notion of being an American and to why people are flocking toward it.

Previous development approaches that have ignored sense of place, place attachments, the value of good neighbors, social interactions, a clean environment, and wild places are of limited value if we want to understand the hopes of those seeking a higher quality of life. We need to move away from a myopic view where the emphasis is on allocating the scarce resource of money and consider how we should spend the scarcer resources of our time and the types of places and environments we want to live in. This kind of development theory better represents the hopes and desires of people, because many activities in our lives cannot be ameliorated by dollars.

A development theory built around hope may seem naive and unrealistic, but it taps a basic aspiration of people, whether in cities or in wilderness areas—to have the ability to live where and how we want.

## Chapter Nine

# "It's My West, Not Yours"

The public lands of the West belong to all Americans, yet it is not unusual or startling to hear, read, or see that "real Westerners" feel that they are getting a raw deal. In their eyes there is a War on the West that is threatening their culture and way of life. The very landscape upon which they depend is being pulled out from under them. The major culprit is the federal government and its control of the land upon which they depend.

In the smaller and more remote towns of the West, a conversation will often drift toward how unfair it is that the federal government has so much control over a person's life. I have had similar conversations in academia as well. The sympathies are with the people and against the large unseen government back east. The rhetoric reaches a crescendo as claims are made that there is a War on the West. This war is forced upon the West not by foreign invaders, but by the peoples' government. Ranchers leading the resistance make it onto the cover of *Time* magazine.[1] They are fighting to preserve the values of the West.

Threats fly in the air, but by and large there are no shootouts down at the corral. The federal troops do not ride into town to put down the rebellion.[2] Instead, the federales, represented by Forest Service and other land managers, are given orders to keep their guns sheathed and to walk lightly. Legal charges fill the air, and local communities claim they have the power to decide how federal lands should be managed.

The federal land managers, who provided the jobs on public lands for people in local communities and land subsidies for ranchers, are vilified for doing their jobs. The very same people who are accused of being too cozy with the people they regulate are portrayed as government storm troopers as they try to see that government regulations are enforced. I say "try," because often they have not been enforced as well as they should have been.

The cows are left to linger on their alloted federal lands a few weeks longer than they should, grazing them more heavily than they should and polluting the waterways. It is easier not to force the issue. There are any number of such regulations that can easily slip through the cracks in a small town rural environment. I am not talking about bribes being given to federal land managers to look the other way, but rather about a way of life where the rangers and their supervisors are sympathetic to the viewpoints of the ranchers, loggers, miners, and others who make a living by using public lands.

This relationship has started to break down as the environmental consequences of looking the other way have become more apparent. And local residents, environmentalists, recreationists, and others started complaining, asking for more oversight. The agencies saw themselves defending unpopular positions in the press. New rangers and land managers, whose thinking was shaped in the crucible of environmentalism, want to see the laws and regulations better enforced, and they have been exerting pressure from within.

Even during the Reagan Administration, caught up in the spirit of environmentalism or for the sake of public relations, heads of agencies such as the Bureau of Land Management declared that "we are the new kid on the block" and will have the best environmental managers in the country.[3] More recently, Interior Secretary Bruce Babbitt's attempt to reform the system and promote the enforcement of environmental restrictions on public land use have made him the heavy, the principal general leading the War on the West.

When a "letter-of-the-law" land manager arrives on the scene, she can get into hot water quickly by enforcing the law too strictly. She is being unreasonable and the "good old boys" demand that she learn what local traditions allow. If not, verbal threats are made, as well as calls demanding that she be transferred.

The West is dotted with towns where various versions of this scenario are played out. The misdeed can be cows overgrazing, violations of timber harvesting, poaching, the shooting of wolves, or polluting streambeds as a result of dredging for gold. When challenged, some people react as if their rights have been violated. There are bars in many of these Western towns where the forest ranger is not welcome. Just as people in a large city must do, the land manager in a two- or three-bar small town must carefully choose a comfortable place to have an after-work drink.

Life for the land manager is more complicated in rural America. It is not like two opposing lawyers in the city who, after the legal wars are over, have a drink together, or the political adversaries who demonize on the floor of Congress each other's proposals and their impact on America and, after the vote, go out on a date. Instead, the forest manager now risks not only verbal abuse and threats but also getting shot at or killed. Enduring these kinds of hazards, is clearly not part of the job description, nor an attractive incentive for people to work in the great outdoors while managing thousands or millions of acres of public wildlands.

For outsiders, this extreme hostility is hard to understand. After all, we all live according to a set of both local and societal rules, regulations, and laws that provide restrictions on what we can do without suffering some potential punishment. We may not agree with all of them, but mostly we abide by them. The difference is that some people in the West see that their way of making a living is being affected. They are a small minority of the people, but they get a lot of attention because they invoke the mythical West, which in their view is being suffocated and forced out of existence.

These rebellions against the federal government ownership are not new; they come and go in cycles. They have been modernized in terms of the techniques used, but remain largely political in nature, using fax machines more often than bulldozers, bombs, or guns, though the potential for violence remains real and should not be discounted. Today they go by the name of the Wise Use Movement, but until recently they were know as Sagebrush Rebellions.

Ronald Reagan, as president, wanting to show his solidarity with the movement, declared himself a Sagebrusher and appointed one as his Interior Secretary. The last Sagebrush Rebellion had started in the late 1970s and received a sympathetic hearing and support from the Reagan administration. Plans were made to transfer some federal lands to the states and to privatize approximately 35 million acres of land. These efforts ultimately fell of their own weight as some of the Western states began asking how they could afford to pay for managing these lands. Environmentalists also mounted campaigns and were successful in recruiting many new vocal members who opposed any attempt to privatize the public lands.

There have been four well-documented Sagebrush Rebellions during our history.[4] They all have been associated with some federal initiative of how federal lands should be managed, and with a Western rebellion against these initiatives. The most recent rebellion and its successors was against preserving lands as wilderness and implementing environmental restrictions on our public lands. The earlier rebellions were over irrigation lands, forest lands, and grazing lands.

## THE WISE USE MOVEMENT

The Wise Use Movement has tried to distance itself from the Sagebrush Rebellion, but continues in its traditions, recruiting adherents from the angry people of the West, declaring environmentalists as the enemy, and wanting to open up all public lands to development via the doctrine of multiple use.[5] It claims to represent the traditional Western way of life, or one view of it anyway. It has gotten a lot of attention in the public media. More than it deserves.

The Wise Use Movement and its backers have tapped into the discontent of a small portion of the American West. What makes its claims powerful is that this portion of the West is where our myths lie. The noble cowboy of movie Westerns continues to ride and herd cattle, the virtuous freedom-seeking lonely miner still hopes to strike it rich, and the modern-day logger with his chain saw, hard hat, or other protective gear required by onerous government regulations earns his living by working in one of the most dangerous and unstable occupations.

The Wise Use Movement claims to speak for the little guy—the fella down the street—who is trying to make a living, to hang on in the rural West. His way of life is threatened by environmentalists who would wipe him and others like him, and their jobs, from the face of the earth. In an interesting twist, one of the leaders of the Wise Use Movement, a freedom-loving free marketeer who has absorbed the contents of the twenty plus volumes of Lenin's works, sees the conflict as one of "class warfare," with the noble Western workers on one side and the environmentalists on the other. The environmentalists and their allies in big government are the enemy.

It all makes for a good show. A more critical, or even a superficial, look at the movement lays bare some of its hyperbole about its being a grass-roots movement supported by lots of small donations from increasingly disenfranchised rural Westerners. The movement probably does receive many small donations, but critics have shown that most of the money comes from the companies with the most to lose if public lands are protected.

The familiar line-up of large corporate giants in the timber, mining, oil and gas, ranching, and other extractive industries have placed their support behind the little guys portrayed in the popular press. A cynic might suspect that their real interest is in corporate profits and maintaining the subsidies that are threatened by public land reforms and the enforcement of environmental laws. The companies have turned over a new leaf. The history of deserting places throughout the West and leaving workers stranded in 1886 or 1996 is relevant only to the most cynical observer.

The Wise Use Movement ignores history and places the blame for past and current loss of jobs almost totally on the shoulders of those "dammed environmentalists." They and environmental regulations are responsible for 90, 95, or 99 percent of all the resource job losses in the American West. False or not, the message hits a nerve, and plays well in various parts of the West. It provides a place to focus anger, place blame, and vindicate yourself.

Ignore the impact of technology, labor productivity increases, global markets, competition from other regions and countries, shifting of jobs outside the United States as a way of reducing labor costs, or a host of other factors that might come into play. I don't need to drive far from my home to be in smaller communities where I can see signs such as "This Business Is Supported By Timber Dollars." The Wise Use Movement taps into the discontent we all feel if our livelihood is threatened. Anger and hostility are natural emotions. Eliminate the jobs of those of us who are tenured professors, and see how we will squeal about the injustice of it all.[6]

The Wise Use Movement presents itself as representing a "class" of hard-working and disaffected people. In this way it is borrowing from the tactics of the civil rights movement and the environmentalists themselves. Blacks are a clearly identifiable group against whom wrongs were and are being carried out. Environmentalists used local organizing and grass-roots techniques to get people outraged about what was happening to local areas and later to national areas. They also sought outside private funding from foundations and increasingly from corporations that want to project a good image.

These earlier movements used campaigns, protests, disruptions, and larger-scale peaceful demonstrations and marches to get their message across, all the time emphasizing the righteousness of their cause. The Wise Use Movement has learned well from its predecessors and has added a few technological twists of its own, using multiple faxes to flood offices, and phone callers to tie up all 80 phone lines into the White House. It is the old strategy of making them think there are more people behind the hill with guns than there really are. Use whatever technique makes the organization seem larger than it really is. Take advantage of mass mailings and media coverage to keep donations rolling in.

The environmental groups were caught off guard. They had put their emphasis on protecting the wilderness and trying to see that laws and regulations on the books were being implemented. They had not paid much attention to what would happen to the people of the West. The Wise Use Movement stepped in with the message "we care, and they don't."

It is a manipulative and cynical approach because most of their corporate backers have no real interest in many of the communities, much less the individual people. The dictates of the market do not allow for such sentiments.[7] Workers whose jobs and livelihood depend on these companies do not feel comfortable in criticizing the hand that gives them their employment, however unstable that may be. It is easier to blame the environmentalists or the federal government. The work ethic remains strong in the rural West, and having environmentalists challenge the value of your work does not go down well in these communities.

The Wise Use Movement appeals to many people in the West who believe that they have an inalienable right to do what they want to do with their property. Anything that restricts that right is a "taking," which in their eyes is forbidden by the Constitution. They have to be fully compensated for any taking that reduces the value of their property.

The Wise Users have embraced this as a central tenet of their movement, wrapping themselves in the flag of property rights and stressing the need to protect themselves from infringement by the federal government. By doing so they have drawn power from the issue of property rights, which is considered sacred in the West. "Property rights" has become the rallying cry of the anti-government movement in the Western states, and is being debated in referendums and state legislatures, as well as being tested in the courts.

The proponents of property rights claim that extreme environmental regulations such as the protection of wetlands limits the right of people to make a profit on their own land. If you can't fill in a wetland, build on it, sell it, and make a profit, then you should be compensated for it. The Endangered Species Act is a major villain to those who see their ability to do what they want with their property as a sacred right.

They fear that the use of their land by any endangered species would result in restrictions that would decrease the value of their property. Their fears have been heightened by a 1995 Supreme Court ruling that the Endangered Species Act can require protecting habitat on private lands. This has spurred on the private property rights advocates to rewrite the Endangered Species Act and pass laws that would require compensation for any losses in property values.[8]

The headquarters of the Wise Use Movement is located in a suburb of Seattle, and in 1995 they were successful in getting a referendum on the ballot to require compensation for declines in property values as a result of any government restrictions. The referendum was voted down primarily because of the huge costs that might accrue to the state. It was seen as a developer's dream. Any restrictions that added to their costs and lowered the subsequent value of their property could require compensation by the taxpayers. Undoubtedly, paying people to develop was a frightening thought to some voters, but it was not the only reason they were alarmed. A significant number of the voters must have realized that owning property is a two-way street. They may own a house and the property upon which it stands and feel a sense of freedom in being able to do what they want with it. There is, however, a responsibility that comes with owning property.

People do not have the right to engage in activities on their property that create harm to other property owners. Polluting your neighbor's property is the most obvious example. In modern urban society there are any number of zoning regulations that restrict incompatible uses of property. Debates in urban areas often focus on determining the point where the level of property restrictions become too onerous and cross the line in restricting freedoms rather than providing protections to property owners as a whole.

There are a number of people in the rural West who say they left the cities to get away from such restrictions, and they don't want big government telling them what they can and cannot do with their land.[9] They seem to feel that these restrictions are new and part of a recent trend that has developed over the last fifty years. Not so.

During the 1630s, when the very first settlers in Massachusetts established towns, they set very specific limits on the use and transfer of private lands. Indeed, the early villages established by settlers from Maine to New York were probably the most comprehensively planned of all American settlements, and in many ways showed a remarkable environmental sensitivity. As one legal observer has put it:

> In American land law, owners have never had the right to engage in unreasonable land uses that cause harm, either to neighboring landowners or the public at large. William Blackstone embraced this idea of limited land ownership: so did James Madison and Thomas Jefferson. And so too do most Americans today.[10]

To some Westerners, harm seems to keep coming from the federal government. This is the conviction of the County Supremacy Movement, which is separate from the Wise Use Movement but which mirrors it, with members often belonging to both. It started in Catron County, New Mexico.

This is a local movement spearheaded by county commissioners who feel that the regulation of federal wildlands is threatening and changing their local way of life. They charge that, since they are affected the most by decisions made about how the public lands will be used, they should have a say in those decisions. And not just a say, but the final say. Catron County, New Mexico drew up an ordinance that has been widely copied by other rural counties in the West.

The Catron County ordinance states that all decisions affecting federal lands in Catron County have to be approved by the local county government. The county commissioners feel it is their responsibility to maintain the local "customs and culture" of the county. Public land decisions affecting that culture must be approved by them. If enforced, this would make forest rangers and other land managers beholden to local politics.

The Catron County ordinance has been used in various parts of the West to try to halt various government actions. In Catron County, the issue was cows and grazing fees. The commissioners claim that requiring ranchers to fence off their cows from streams and raising grazing fees, even slightly, can put some ranchers out of business. Other counties have other issues. In Boundary County, Idaho, which straddles the border with Canada, county commissioners passed the same ordinance. However, the issue was not cows, but worries about the declining tree harvesting in the adjacent national forest.

A rash of these ordinances have been passed to try to change some action(s) considered necessary by the federal land managers, but not by the county commissioners. What they all seem to have in common is that they are unconstitutional. The Boundary County ordinance was the first one tested in court. The judge clearly stated that the county ordinances had no basis in law and violated both the state and the United States Constitution. This was a severe blow for the county commissioners and the property rights crowd who wave their Constitutions at will, claiming that their rights are being violated and that even raising grazing fees is a taking of their property.

The county commissioners, along with local ranchers and others, often claim too that the public lands do not belong to all American citizens. The federal government has no right to them; if anything they belong to the states. They claim that each state that entered the union had the right to claim their own public lands on an equal footing, and that by keeping lands that rightfully should have passed to the states, the federal government is in violation of the Constitution. That the Western states entered the union accepting the ownership of these lands by the federal government is in their eyes a non-binding act. Consequently, they tell land managers that they can't manage land that they don't own. Ultimately, their appeal is emotional, since such legal arguments are not taken seriously by legal scholars.

The ordinances may clearly not be constitutional, but they do have an impact, and a very negative one on federal managers. It takes time to fight the ordinances as they wind their way through the courts and as they will continue to do until the movement peters out. Even if they reach the U.S. Supreme Court, they stand little chance of being upheld. Regardless of whether the justices have been predominantly conservative or liberal, the Supreme Court has consistently supported the federal government in resource conflicts where the issue has been more local control.[11]

A larger purpose, and probably the main purpose, of these ordinances has been to get the attention of the land managers and to intimidate them so that they will hesitate before going forward with some action that local public lands commodity extractors will find harmful to their pocketbooks. Keep the land managers huddled in their offices, not out checking or enforcing land use regulations. The scheme seems to have been successful in places.

The irony of all this is that, in places where these ordinances have been adopted, many of these same people will not complain when their lives are disrupted as a result of decisions made by a private corporation. There are no protests organized when private landowners raise their grazing fees. If a mining company shuts down a mine because the price of the mineral drops and leaves town, well, that is how it goes. If a timber company leaves for similar reasons or because it has cut all the trees it wants, again that is just business. And businesses have to do what is best for their profits. People can understand that. Let the government make similar decisions for the good of the land and, among a small segment of the West, all hell breaks loose.

If all the West were privately owned, the protests would be muted indeed. It is understandable, but somewhat deplorable, that companies would pour money into causes that create great hardships for federal managers for doing what they think is right, when their own managers are isolated from having to face protests for the shattered lives and places they sometimes leave behind.

The visibility of these movements under a Wise Use banner of "let's keep extracting as much as we can from our public lands while we can" creates a false impression of the West. As I have shown earlier, most people in these counties want to protect the public lands, not develop them. The economies of the West are not dependent on farming, ranching, mining, or timber harvesting.

The noise made by the Wise Use Movement represents a small cross-section of people within these commodity groups. It is unfair to stereotype all ranchers or loggers as people who buy into the extreme rhetoric of the Wise Use Movement. Most of the people in the West do not want the environmental laws repealed, as the Wise Use Movement does. Most people want to work toward a protective strategy, with commodity extraction limited to what can be done using the principles of good stewardship, which the agencies are still struggling to define.

The Wise Use and other movements, while not reflective of Western attitudes, are remnants of the Old West, and not the romantic version of it often portrayed in the media. The real historical West is only now learning to be more accurately portrayed.

## WILDERNESS AND HATE

The West is a land of social as well as physical extremes. It is also younger than its historical nemesis, the East. It was over 200 years after the founding of the Massachusetts colonies before the first white settler set foot in Idaho, and he was coming from the West. The interior West attracted some settlers, but very few compared to the land of gold and plenty in California.

Still today the remoteness of parts of the interior West and its wildlands is what attracts various people to the region. The Sagebrush Rebellion and the Wise Use and County Supremacy movements represent the cries of people still not too far removed from the pioneer stage of felling the forests, dredging streambeds for slivers of gold, and running cows with little concern for the impacts on the landscape.

They raise the Constitution as their source of authority. Their politics is fundamental—don't tread on me. For the most part, even though they demonstrate and carry guns on their hips or in a pickup rack, they don't go beyond verbal threats. Most of their activity consists of filing legal briefs and lawsuits and looking for sympathetic judges.

The threat of violence is made clearer by the more extreme cousins of these property right constitutionalists, those who come west to hide out or to hate. Places in the interior West have always been attractive to survivalists, people who want to get physically away from the mainstream and drop out, either living off of the land, or waiting and preparing themselves for Armageddon.

More recently they have been joined by groups that want to live surrounded by other like-minded righteous persons in Christian communities, that want to form a rural communal suburb, or that want to live more nefariously in more rural militaristic surroundings where they will defend the Constitution from attacks from their own government. I see examples of such paranoia on my own campus in bumper stickers that read "I Love My Country. It's My Government I Can't Stand."

People have a right to live where they want, and to think whatever they want. Usually, they don't pose a real threat to their neighbors. The neo-Nazis, however, are not simply survivalists biding their time, expecting the worst. They have their own plan, setting up an all-white empire in the Northwest.

What all of these groups—from people hiding out and surviving to neo-Nazis—have in common is the use of wilderness and an image of a wilder West either as a place to drop out or to provide a form of protection from the rest of society. The conventional view of our wild public lands is that they are a place to retreat to for mental and physical renewal, or to seek a spiritual reawakening.

The public wildlands are used primarily by hikers, backpackers, exhausted corporate executives, and a variety of people seeking either solace or recreation. Recreationalists may differ widely in how they would like to see public wildlands used, from cross-country skiing or snowmobiling to casting for wild trout or riding jet boats down the canyon rivers. These views of how the lands should be used may differ, but they are all very different from the view that regards public lands as a buffer against outsiders.

Living adjacent to or near federal wildlands in order to keep out the rest of American society is the choice of only a relatively small number of people who get an inordinate amount of attention and coverage, perhaps because they are really threatening and dangerous. Certainly, the idea of neo-Nazis in the Northwest is odious to most Americans, and frightening as well. Idaho has gotten a lot of media attention for its small contingent of neo-Nazis in the northern part of the state.

A feature article in the Sunday *New York Times Magazine* was titled "Enter Government Hating, Home Schooling, Scripture Quoting Idaho, the New Leave-Me-Alone America at Its Most Extreme."[12] This is certainly not how the state likes to see itself portrayed. The Governor railed about the unfairness of it. The neo-Nazis have brought a notoriety to rural places and states. Is this negative attention unfair? In many ways it is, although by taking a passive role toward these fringe groups, the communities and states allow the impression to linger that they just might agree with some of the principles these hate groups espouse.

Hate groups get a lot more attention in the rural West because they stand out more and are more representative of the local populations in that they are all primarily white. The cities these hate mongers leave behind are much more culturally diverse. Neo-Nazis exist, of course, in cities as well as in the rural West. The diversity of cities means that it is difficult to argue that, because they contain neo-Nazis, the city itself is a haven for such people.

When I lived in the City of Chicago during the early 1970s, neo-Nazis were marching in the streets protesting the movement of blacks out of the ghetto and into white neighborhoods in West Chicago. While there were some people in Chicago who probably sympathized with the neo-Nazis and the presence of neo-Nazis was a problem for the city, nobody suggested that Chicago was a city of neo-Nazis. Chicago is a city that is highly diverse racially and ethnically and is still geographically segregated by large neighborhoods and community groupings. The neo-Nazis were a fringe group trying to take advantage of white fears generated by the movement of blacks who are trying to improve their lot in life by moving out of ghetto areas, just as white immigrant groups had done before them.

I grew up in an older suburb just outside New York City. During the summers of my college years, I worked for a moving company carrying washing machines and furniture in and out of apartments in the Bronx and other boroughs of the city and the surrounding area. I worked with a lot of different people who were carrying furniture. One of them was a fellow who had been given the equivalent of a dishonorable discharge from the Marines, as he was proud of pointing out. We didn't talk much about politics as I recall.

After one summer, when I was back in college I learned that ex-Marine had gone into the offices and warehouse of the moving company and opened fire with automatic weapons, killing about five people. The police arrived and shot and killed him. When they went to his apartment, they found it full of Nazi artifacts and flags. He belonged to a neo-Nazi group but, as far as I knew from my own experience, kept quiet about it, except for the usual "blame the government for what is wrong with America" talk. Neither my hometown of New Rochelle nor the greater New York City area have been considered to be neo-Nazi strongholds despite the presence of these groups in their midst. Not so for parts of the West.

There is a tendency to identify parts of the West with these hate groups. It may be a continuation of the stereotyping that portrays the anti-government, wise use, and county-supremacy activists and sympathizers as representatives of the true West. The current demographic trends of the people moving into the inner West also present a picture of white migrants seeking surroundings that are less diverse. Consider recent migrants to counties containing federal wilderness. The people moving to wilderness counties are almost all white, and the counties they are moving into have a population that is over 99 percent white.[13] Obviously, groups planning to build white empires prefer to move to places that are already white. Whether the other persons moving in, the overwhelming majority (which, I hope we can safely say, doesn't quietly share the same values as the hate groups), share some of the same motivations for moving there is an open question.

Many people do say they are moving to leave behind cities with their crime and social ills. What surveys normally don't ask is whether they leave because of their racial and ethnic characteristics.[14] The suspicion that people do move to rural areas to avoid certain racial and ethnic groups seems quite plausible. The areas offer a certain lifestyle and ambiance, predominantly white and with politically conservative values. The rural West provides space that protects against unwanted social proximity.

The recent movement of some people into a largely white interior Northwest fleeing other more culturally diverse places is reminiscent of an earlier group of people fleeing to the region and bringing their prejudices and intolerance with them. For example, the earlier migration brought numbers of Confederate refugees, who had fought to preserve slavery, into what was then the remote Idaho Territory, which was established in 1863.

These new migrants found and welded considerable political power. They also found another group, the Chinese, who had come to work in the mines. The Chinese quickly bore the brunt of racism and intolerance. Anti-Chinese organizations and political parties sprang up and were successful in passing a constitutional provision in 1889 disqualifying from voting or holding office any persons who were "foreign-born of Chinese and Mongolian descent." It was not uncommon for Chinese to be killed or lynched in the Territory.

The attitudes towards the Chinese were replicated in intolerance towards the Mormons, contempt of the rights of Indians, as well as of Blacks and Hispanics. This historical intolerance may reflect the inability, as some argue, of the current Idaho majority to accept cultural differences. Indeed, it has been argued that the Northwest may be becoming less tolerant of diversity.[15]

The settling of the West with "refugees" from the southern states and intolerant individualistic "pioneers" may have been coincidental, a historical event of little significance for present-day events. I leave that for historians to debate. Nonetheless, I have found on occasion a looseness of tongue that includes racial, ethnic, and religious jokes and slurs in both private and public places that is reminiscent of an older South. Politicians from this white inner Northwest who talk of the "endangered" white males beset by government rules and of possible loss of their property rights risk being perceived or labeled as racist, just as were some of the defenders of segregation and states' rights in the South.

Labeling is inherently unfair, but unfair or not, part of the burden is on the people of the West to show that these stereotypes fit only a small lunatic fringe. And when places are directly challenged with the presence of neo-Nazi activity, places such as Coeur d'Alene, Idaho, or Billings, Montana, have responded strongly, even getting international awards and recognition for their efforts. Yet it is the actions of normal people that set the tone and create an atmosphere of tolerance—"live and let live"—that the West is famous for or that allow the historical prejudices to keep reasserting themselves by excluding or making outsiders feel uncomfortable.

I enjoy going into the cafes of the rural West and have always taken for granted the hospitality and interesting conversations that I have had while there. It saddens me when I talk to people of other racial groups or in mixed marriages who point out that they never stop in these cafes because of the hostility that they encounter. Instead, when traveling, they only stop in the larger gas stations between the larger cities in the region. When asked if this is unique to the West, they often reply that it is not. They would not feel comfortable stopping in rural towns in general, especially in the West or the South. The degree of prejudice that minority groups, whether Black, Hispanic, or Indian, feel in any rural area may not vary significantly, but it is present and to ignore it presents a more idealized picture of western rural life than actually exists.

An indication of the pervasiveness of prejudice is the establishment by the United Methodist Church of a task force on rural racism. To its distress, the results of hearings held throughout the country were that examples of rural racism are easy to find in all regions of the country. The task force concluded that racism continues to be a pervasive force within both the church and rural communities; that people and churches ignore or hide from racism; and that silence contributes to continued racism.[16] Such racism exists and continues to exist in the fabric of American society, according to noted historian Arthur Schlesinger, Jr.:

> As for the nonwhite peoples—those long in America whom the Europeans overran and massacred, or those hauled in against their will from Africa and Asia—deeply bred racism put them all, red Americans, black Americans, yellow Americans, brown Americans, well outside the pale. The curse of racism was the great failure of the American experiment, the glaring contradiction of American ideals and the *still crippling desease* (italics mine) of American life.[17]

The white supremacists groups continue to exploit anger and search for scapegoats in localized depressed rural economies throughout America. It is no accident that recruiting efforts of both the Wise Use Movement and the white supremacists are in the declining mining, timber, and ranching sectors of western communities. It is easier to tap into the frustration of people who are suffering the most in the changing economies of the West. In the Northwest, where the Native Americans are confined to reservations, the remaining population is white, especially in the more rural parts of the region.

The very concentration of whites in the region as a whole attracts white supremacist/patriot groups to the area. Although there has been a trend towards greater population diversity in the region, the area still remains considerably less diverse than the West as a whole. For example, in the Northwest in 1970, in only 7 counties was 5 percent of the population nonwhite (Figure 9.1). By 1990, this had increased to 30 counties, but, if the regional average for the West is used, there are only 2 such counties. The region is slowly becoming more culturally diverse, although still much less so than the rest of the region and the United States as a whole.

The history, relative remoteness, and continued relative "whiteness" of the inner West continues to make it attractive to militia and "patriot" groups. The anti-government sentiments are at the heart of the militia groups, which would like to set up their own constitutional governments that more likely would become the American "dictatorships of the people." The "patriots" would turn back the historical clock by not allowing women to vote—allowing that privilege only to white Christian men. As in the past, Blacks, Asians, Hispanics, and any other minorities would not be allowed to be citizens of their "nation."[18]

Like immigrants in the past, some of these "patriots" headed west. Today there are estimated to be a total of more than 400 militia groups; they can be found in each and every state in the nation. Michigan (50), Florida (43), and California (34) are distinguished by having the greatest number of these militia and "patriot" groups.

Most of the people in these groups stay in their home states, but those who move into the rural West draw a considerable amount of attention, especially given the publicity and aftermath of the Ruby

Counties with greater than 5% non-white,
non-native American residents

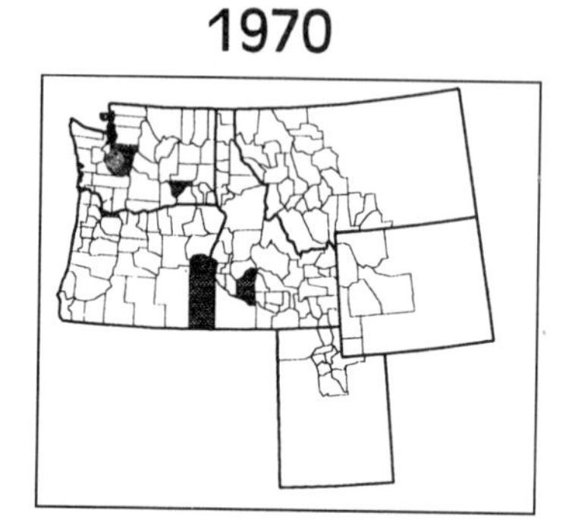


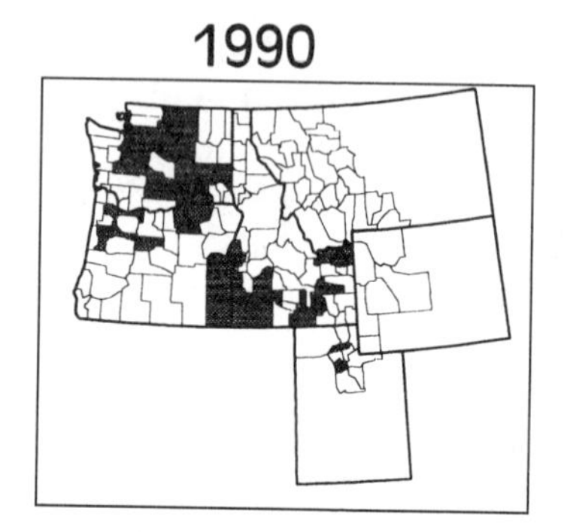


Counties in 1990 with greater than 21% non-white,
non-native American residents*

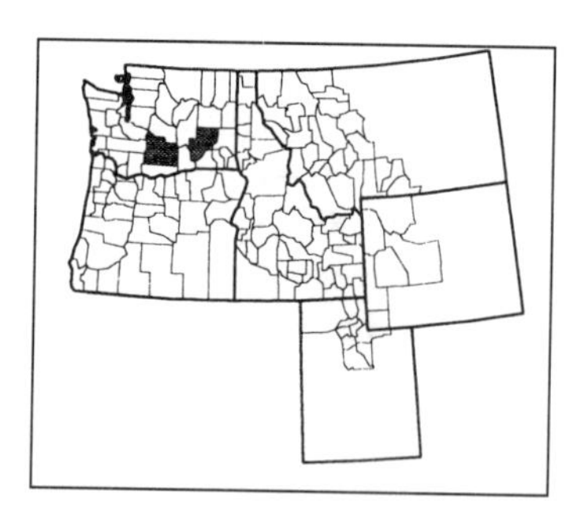

*21% was the regional average for 1990
for the 11-state "West Region," including
CA,AZ, NM, CO, NV, OR, WA, ID, UT, MT, and WY

***Figure 9.1*** Nonwhite Population in the Northwest. *Source:* Based on various U.S. Bureau of Census estimates.

Ridge siege in Idaho, the standoff of the Freeman in Eastern Montana, and the capture of a suspect in the Unibomber case who was living as a recluse in western Montana.

Even though the western states have a small number of these groups (Idaho, 4; Montana, 4; Wyoming, 2) compared to other states, these recent incidents portray a region that journalists no longer call the "Wild West" but, rather, the "Wacky West." People wonder whether the West is "full of kooks," and self-deprecating locals in Montana say, "at least the cows are sane" and that it is "the last best place to hide."

Though people may argue just which communities and states are the most intolerant and welcoming of the over 800 militias, white supremacy, common-law courts, and other groups that comprise the "patriot" movement, there is no reason to believe that one place or region will become the "homeland" or the new "white fatherland," as some of the leaders of these groups would like parts of the West to become. What the western states do have is a landscape that is easier to hide in, fewer neighbors, and a culture of guns and weak governmental structures to confront such groups. Whether they have a population that is more willing to accept the propaganda of these "patriot" groups is open to debate.

The town of Kamiah, Idaho, with a population just over 1,000, has become the designated site of "Almost Heaven," a new "patriot covenant community." This 200-acre development is to be a place where white Christians can come and live among similarly minded people. They will not, as their leaders say, look for trouble, but they will be armed and prepared should any trouble come to them. There have been fears by some of the locals that they will not obey the laws and will try to take over the local government.

The people of Kamiah have been split by the imminent arrival of this new "survivalist-patriot" group. Some feel threatened by an influx of people who are said to be intolerant of others and who are possibly preparing for a confrontation with the much maligned "big brother" government. Others are more welcoming, or at least more tolerant, taking a "let's-wait-and-see" attitude. Town meetings have been held with the leaders of this group, who assure everybody that they just want to be "good neighbors." Ironically, people looking for a community of like-minded people will find that they are moving into Indian country and that there is a small Quaker community in the area as well.

The tensions building in Kamiah as a result of its being labeled a "home for patriots" have not been eased by a demonstration by students who objected to not celebrating the Martin Luther King, Jr., holiday. This brought out further complaints by the Native American population that their needs were not being considered in the local school system.

Kamiah is a community through which I have passed and stopped frequently, since it is not far from the Selway-Bitterroot Wilderness Area, one of the most beautiful areas in Idaho and America. It is a community that traditionally has been dependent on the timber industry but that has tried to diversify by concentrating on tourism. Located on the scenic Lewis-Clark highway by the Clearwater River, it has tried to divert people off the road and into its small downtown by building on the image of the old West. The downtown stores have been covered with fake fronts to make it look more like a town of the mythical West than the actual "real" western community that it is. The effect is jolting, and I hope, for its sake, successful. It has an air of artificiality and replaces the realism of a town rooted in a working past with one of the images of a frontier garnered from dime store cowboy novels and movies. This has worked for a few towns in the West, and, obviously, the citizens of Kamiah are hoping it will work for them. Whether the presence of the development Almost Heaven becomes a greater draw, though perhaps of a different sort of tourism, remains to be seen.

The neo-Nazi groups are not alone in using that space. Although they do not necessarily use the same rhetoric, the constitutional militias have been sprouting up in the rural West, shouting familiar anti-government slogans and declaring that they need to be ready to defend themselves. These marginal groups are not new, and are just as prevalent in Michigan and Indiana as in Montana or Idaho. What gave them greater visibility was the political climate in the West, where they became part of the anti-government movement.

Western congresswomen and senators began giving speeches in which they stressed that, although they did not believe in taking up arms against the federal government, they understood the frustrations that were driving people to feel the way they did. Then came the Oklahoma City bombing of the federal building by suspects tied to the militia movement.

The bombing of a federal building, with the loss of life of men, women, and children, cast doubt on the notion that the people who caused it were ordinary Americans estranged from their government. Recent evidence shows that, after the bombings, three national militia networks began campaigns calling for the surveillance of government buildings and the offices of civil rights organizations. The Aryan Nation World Congress held one of its largest meetings in northern Idaho, and called for the merger of the neo-Nazis with the militia groups and the stockpiling of weapons in preparation for an impending race war.[19]

The movement of groups that want to overthrow the government or that are waiting for their government to attack them into public lands counties, puts a chill on the perception of the West as the land of opportunity. The sight of ranchers and county commissioners threatening forest rangers and other land managers because of their efforts to keep the lands from being further degraded dashes to pieces the myth of the self-reliant Westerner on the side of virtue and justice. The tawdriness of the tactics used and the company some of these people and groups keep tarnish the image most people want to have of people in the American West.

The reaction of the federal land agencies to the threats and actions against their scientists and land managers has been of a subdued, "keep a low profile, don't anger the local citizens" variety. Even when the offices of the local forest ranger have been bombed, there have been no strong words from headquarters, much less any actions taken, legal or otherwise. The cautious response has been to travel in pairs to avoid confrontations, and, if confronted, to retreat. From one perspective, this restraint appears commendable; from another perspective, it appears shameful, as it leaves land managers vulnerable and emboldens others to continue harassment as these federal employees try to implement the law and their duties under trying conditions.

The presence of people and militia expecting to be attacked by their own government can lead to local confrontations, many of which don't get reported to the local police or media. For example, just this year, north of where I live, there were two linemen who were up on the poles going about their business when two "patriot" types passing by assumed that what the linemen were doing was attaching equipment to spy on them in order to send in helicopters to attack them. So they opened fire on these exposed linemen, who naturally hustled down the telephone poles as quickly as possible and ran to their truck. It happened that they had weapons in their truck—not to defend themselves, but in case they saw a deer (since it was hunting season). They shot back at these "patriots," who then calmed down, and they all had a talk. It turned out that the "patriots" were new to the area, so the linemen explained in colorful terms a bit about western etiquette. Needless to say, such incidents don't make these new residents popular with the "locals."

## THE URBAN AND RURAL WEST

The rancor in the rural West is also directed toward the urban West. The people in rural areas see themselves as part of the real West, whereas urban areas are in danger of becoming more and more like urban areas everywhere else. There is also the threat throughout much of the region that there will be a battle over who will get the available water and what it will be used for. The battles in rural areas that pit ranchers, foresters, and other commodity users against not only land managers but against the larger urban public raises the question of whether the West will break into two contending parts, urban and rural, that will constantly squabble over who gets to decide how the public wildlands will be used.

There are any number of battles that appear to divide along the rural-urban line. When people in urban areas use the public lands, they do so mainly for recreational or aesthetic reasons. For them, the landscape is not simply a commodity to be consumed. They outnumber by far the people who live in rural areas and either extract some type of commodity from the landscape or work for corporations that do. There is a tendency to somehow discount the way people in the urban areas and larger towns and cities that are not dependent on resource extraction feel about and value the wild landscape.

There is a lot of discussion about how federal wildlands management needs to be done with people being considered as part of the larger ecosystem and with an awareness that human needs cannot be overlooked. One reason the various property rights and county supremacy movements are popular in many small towns is the perception of environmentalists as the enemy. The larger environmental groups and their more regionally based local cousins have not been very successful in considering the future needs of the people in these smaller Western communities. The emphasis has been, and in many cases continues to be, almost solely on protecting large areas of public wilderness. Recognizing that the land management policies of the past have failed, those who feel a need to make a fresh start concentrate on discussions of landscape reconstruction and pay little attention to what happens to the people currently benefiting from the failed policies.

Offended by this lack of consideration, a diverse group of rural people are driven into the arms of fringe groups that make many promises and provide release for anger directed at an easily identifiable enemy. That most of their promises will prove to be false, and their notion that today's problems can be solved by going back to the "good old days" quite unrealistic, may be beside the point.

A symbolic rallying cry against an external enemy provides some satisfaction. But like the people of the South who resisted the civil rights movement, asserting constitutional and states rights as a rationalization for continuing what the rest of the country has already seen to be no longer sustainable as a social system, they must know in their hearts as did many Southerners, that the old ways are wrong and doomed.

Environmentalists may use hyperbole as well as science to promote their cause, but the landscape of the West does not lie. The satellites in the sky show what has happened and what is happening not only to the Brazilian rain forests, but to the American West's public wilderness as well. The facts slowly build up. Identifying with reactionary forces that deny the truth about what has been happening to the landscape will not make the people of the West proud as they look back in the years to come.

Some environmental organizations have recognized that ignoring the people in the immediate vicinity of the public wildlands has been a mistake. There are efforts to reach out to small communities and to help them consider how they would like to adjust to the changes taking place around them as well as to larger global forces that are affecting them as well.

These efforts involve going out and working in communities and showing them what can be done and what is being done. In concrete terms, this means that some environmentalists are saying that they agree that not all logging is bad. It is important, too, to find individuals and companies that are involved in logging, mining, ranching or other activities in ways that are both environmentally sound and profitable. By providing positive examples and by working together to increase the number of environmentally responsible operations, conflicts can hopefully be diminished and a better future assured for these communities.[20]

Efforts such as these are to be commended. The rural communities do provide a part of the heart of the West. The tendency to pay less attention to the needs of the urban population is not a healthy one. Take the case of the Utah wilderness. Along with some of their constituents and the help of other politicians, county commissioners tried to railroad through a very limited Utah Wilderness Bill, even though most of the other people in Utah preferred that much more federal wilderness be designated.

Utah wilderness—a desert landscape unlike anywhere else, the real landscape behind our cowboy western movies—became a national cause. Despite great political pressures to open most of it for development, people in the rest of the nation joined in to demand that more of what is part of their national landscape be preserved. In such a context, allowing a few hundred people and their moneyed backers to decide how federal lands should be managed is unwise and unjust. That is the dilemma of the urban and rural West.

The urban people of the West are not clamoring that their constitutional rights are being violated, that federal wildlands should be given back to the states, and that land managers should be kicked off the lands. The federal lands are an important reason for why they choose to live in the urban areas or smaller cities throughout the West. They want to have as much say as their rural neighbors in how these lands are used. In many cases they may actually use these lands more than rural residents. Their needs, however, are more highly discounted by politicians and land managers alike. Land managers are more easily influenced by their more immediate rural neighbors than the more distant city dwellers.

This polarization between rural and urban people can be carried to extremes. And it is the extremes that are portrayed in the media. The latté-drinking, mountain-bike-riding, skiing urbanite, on the one hand, and the beer-drinking ruralite with his pickup truck, rifle, or handgun, on the other hand, meet on the edge of our wilderness, itching for a showdown. This is a caricature that either is already inaccurate or soon will be because rural places are changing rapidly in the West.

In a survey of how people felt our public lands should be managed, I was surprised at the degree of similarity between rural and urban people. Both urban and rural people gave as the top priority of wildlands management the protection of the watershed and ecosystems. The media and western politicians would have you think that the rural people would rank the extractive activities highest, especially timber harvesting and ranching.[21]

This is encouraging, because it questions the validity of another myth about the West, that there are vast differences between people depending on where they live. The degree of agreement about public land management is not as surprising as it might be when we remember that most people in the West are not tied to cutting trees or herding cows, but care about the "health" of their public wildlands. If it was presented to the voters directly instead of filtered through politicians dependent on political action group contributions, the message to protect our lands would be much stronger and clearer.

The real West may not be rooted in the rural areas around which the "it is my place, not yours" Western fringe crowd lives. It may be in the smaller cities such as Moscow or Bozeman, or any number of other places. Wallace Stegner recognized it first, and said it better. He said that if he were advising a documentary film-maker who was doing a fifty-six minute film on the quintessential West:

> I would steer him away from broken-down rodeo riders, away from the towns of the energy boom, away from the cities, and send him to just such a little city as Missoula or Corvallis, some settlement that has managed against difficulty to make itself into a place and is likely to remain one . . . these towns and cities still close to the earth, intimate and interdependent in their shared community, shared optimism, and shared memory. These are the seedbeds of an emergent western culture. They are likely to be there when the agribusiness fields have turned to alkali flats and the dams have silted up, when the waves of overpopulation have receded, leaving the stickers to get on with the business of adaption.[22]

I might add that he probably would steer them away from the towns of the wise users and county supremacists, to say nothing of the militia and neo-Nazis. They have not proven themselves to be the stickers or the lovers of the Western landscape.

All of the human components of the West are moving toward a meeting, one that Wallace Stegner seemed to regret, as he indicated in his comments about overpopulation washing over the landscape. The western landscape, like other regions, is pockmarked with a hierarchy of hamlets, towns, and small and large cities. They are all beneficiaries of the physical and existential space of the West. As these places grow (few are declining now), their demands on the public lands will grow, and nowhere is this more evident than in the Pacific Northwest.

## THE NORTH CASCADES: URBAN MEETS RURAL

The North Cascades is the site of a visionary proposal that encapsulates the possibility of making a transition from a commodity-based management strategy to a more holistic ecosystem management of our public lands. If implemented, it could serve as a model for the rest of the world. The proposal crosses the border and would establish a Cascades International Park that would be cooperatively managed by U.S. and Canadian land managers.

The new international park would be part of a Greater North Cascades Ecosystem (Figure 9.2) and would be surrounded by public lands where extractive and other activities could take place if they did not harm the ecosystem.[23] The U.S. side of the proposed international park has the bulk of the land and it contains North Cascades National Park, and Ross Lake and Lake Chelan National Recreation areas, all managed by the National Park Service. It also contains eight wilderness areas managed either by the Park Service or the Forest Service, as well as several National Forests. The Canadian side presently has two provincial parks, recreation areas, and several other protected and restoration areas.

The proposal for an international park is not new. It originally was proposed in 1920, but not much happened with it. Over the last ten years various local groups have been pushing for the park, and

*Figure 9.2* Greater North Cascades Ecosystem. *Source:* Adapted from Northwest Ecosystems Alliance.

it has been gathering increasing support from local people, politicians, and the Secretary of the Interior. The establishment of the park itself is a worthy goal, but the proposal to surround the park with other federal lands designated for ecosystem management goes even further.

Most of this land is Forest Service land currently managed for timber production and recreation. It contains the infamous Pinchot National Forest, which stands as a monument to the impact of clear-cutting. Much of the Forest Service's logged land would be rehabilitated. These lands would be part of an ecosystem zone where primary emphasis would be given to protecting the ecosystem. Logging and other activities would only be allowed if they did no harm. On Forest Service lands, the priority would be exactly the opposite of current management priorities. The same would be true for recreation, which would be allowed as long as it did not overwhelm the natural system.[24] And recreational uses will clearly increase in the years to come.

The proposed greater ecosystem management area, with the park and wilderness at its core, is located in one of the fastest-growing areas in the country. The North Cascades ecosystem is located on the edge of what I have called the "real" West. On a climatic and vegetation basis, clearly it is not a part of the arid West because it extends westward to the Pacific. It lies between a continually advancing urbanization at the western edge and an interior that traditionally has been associated with a resource extractive West.

Seattle has been attracting large numbers of migrants during the 1980s and 1990s, and has been rated the best city to live in the United States. As we proceed eastward into the interior, we begin to encounter the current image of the reactionary West, with its county supremacists, militias, and neo-Nazis. Indeed, counties within the eastern slopes of the proposed park ecosystem have passed their versions of the Catron County ordinance asserting that they should have control over how these lands are managed. County movements exist on the Canadian side as well. These forces do not like the idea of an international park, as I found out.

There have been no agreements made, no designations of land changes, no bills drafted or laws passed. The park and its surrounding ecosystem remain an idea whose time is yet to come. The various organizations and agencies on both sides of the border organized a conference to discuss the desirability of the park and the issues involved in implementing it. I was one of the speakers invited to address about 250 interested private and public persons on the feasibility of their vision.

The conference was held on the campus of the University of Washington. As I approached the building where the conference was being held, I noticed a commotion outside it. A group of people were walking around holding signs and posters and chanting "No International Park" and other slogans. The other slogans referred to the loss of jobs and property rights that the proposed park would cause. The Wise Use Movement had organized a protest against the conference and the park. The demonstrators were in high spirits and were having a good time as representatives from the local news media talked to and videotaped them for the evening news.

I felt a bit exhilarated myself. Not many places where I give talks get picketed. It also brought home how well prepared the Wise Use Movement is to organize against any activities that threaten to change the dominant extractive uses of our federal wildlands. I had thought that setting up an international park with our friendly neighbor to the north not only would be an obvious step in the right direction, but a non-controversial one as well.

The demonstrators portrayed it as a means of throwing people who live around the proposed park out of work, and others spoke of ultimate control of the park by the United Nations, with some even suggesting that helicopters would fly over the area to keep people out. All of this was nonsense, but it seemed to make perfectly good sense to the true believers. The discussions were in the formative stages, and there were no plans at all to include any private lands in the park or in the surrounding federal lands.

There would be potential decreases in the timber harvest, which might affect some jobs, but that was not the major issue of the demonstrators and other critics. If anything, the growing population of Seattle and the spread of population growth on both sides of the Cascades would put more recreational demands on the region. Counties around the present national park/forest lands in the North Cascades had a population increase of 22 percent during the 1980s, while other counties not in proximity to federal wildlands in the state had population increases of only half that.

Establishing an international park would only increase the use of the area. If anything, job growth would be faster than without an expanded park and surrounding ecosystem. Some of the wise users and county commissioners from both sides of the border were paid to attend the conference and participated in the frank, lively, but reasoned discussions they helped to precipitate. I spoke with a number of them, and they did not dispute the changes taking place.

The international park and surrounding federally protected ecosystem would be a great improvement over the current fragmented federal management of the North Cascades, which different agencies had cut into different pieces in a haphazard fashion. While the Park Service caters to the growing recreation demand, the Forest Service clear-cuts habitat, and helps to threaten some species with extinction.

The declining salmon are an obvious example that current management practices cannot sustain a healthy North Cascades ecosystem. People in the Forest Service want to have the ability to change past and current practices and manage for the long-term viability of the North Cascades ecosystem. Such people are still constrained by the agencies' cultural dogma and the inertia that thwarts institutional change.

It is ironic that there appears to be more interest in cooperation between the United States and Canada than between the agencies and local people in the area. Cooperation is necessary and will undoubtedly come with time. It is a shame that the process has to contend with charges by a well-funded fringe group that does not back up its charges with evidence, but resorts instead to half-truths or outright lies.

Charges are thrown about that the United Nations will take charge and that concern for other life forms and ecosystems poses a threat as great as communism. But now that communism is in decline, environmental fascism is seen as the latest threat. Fortunately, even though such charges get much publicity, the groups behind them do not represent the Real West, or the New West.

The people of the North Cascades ecosystem could work toward providing an example of how to live in better harmony with their environment even as population increases and disperses throughout the region. They have the opportunity to lead in ways that the rest of the country might follow as we struggle to move away from past management practices that more and more people realize were wrongheaded or destructive.

The alliance of groups promoting the North Cascades Ecoregion is also working toward an economic strategy to assist local communities in the region make any necessary transition. A strategy of involving local people in the development of management practices of the proposed park can only help to reveal the distortions of the more reactionary groups opposing the project. The people who love and respect the West regardless of whether they are newcomers, old-timers, or natives are not afraid to share it with other Americans whatever their beliefs or skin color.

*Chapter Ten*

# Future Directions for Wilderness

The future of federal wildlands policy in the American West can take several directions. One would be to do nothing. Leave the current laws and regulations in place and simply continue to manage, adjusting as necessary to the political winds of the time. Such a scenario is highly unlikely. The need for change has been recognized by all the sides contesting to get their way in determining how our wilderness should be managed or not managed. The rhetoric has moved beyond the simple shrillness of environmental advocacy group fundraising letters predicting doom and gloom unless something is done at once, and earnestly soliciting a donation to the cause. The rise of the Wise Use Movement, which has borrowed some of the organizing tactics and strategies of the environmental movement and added their own unique militaristic and threatening rhetoric about there being a War on the West, has put a greater focus on the rights of individuals and large corporations to earn a living from the landscape of the West, especially if it is subsidized by the citizens of America as a whole.

As regards commodity extraction activities that destroy or greatly modify the underlying ecosystems, there has been less faith in the federal agencies' rhetoric of conservation and multiple use. There is an increasing perception that such management has contributed to a form of individual and corporate welfare, or subsidized workfare, and the granting of subsidies to maintain a Western lifestyle and corporate profits. There need be nothing wrong with this if the rest of the citizens want to provide these subsidies to individuals and corporations.

Many of these subsidies were not an issue in the past, or were justified by a variety of proffered benefits to the rest of America. With more scrutiny and attention to what has been happening to their lands, the rest of America is questioning both the benefits of current management, and whether changes are taking place fast enough. Within the West itself, new migrants are bringing new viewpoints and are lobbying for changes as they move deeper into the American West. All of this suggests that doing nothing remains highly unlikely, and indeed the history of our wildlands has been one of constantly changing demands concerning how they are to be used.

Another scenario is that wildlands policy will continue as it is, with only a bit of tinkering at the edges, or with some adjusting and fine-tuning of management strategies as the need becomes apparent. This is the policy recommended by most observers and the managers themselves. It is an admission that things are not going quite as they should, but with some readjustments here and there the agencies will be more responsive to the public. Nonetheless, the managers remain the "experts." The embracing of ecosystem management as a framework within which to continue multiple-use management with a greater emphasis on both biological and human needs is similar to following a well-worn path.

The continued calls for reorganizing priorities within individual agencies and for inter-agency cooperation is another example of how some changes or agreements can be used to solve problems and improve wildlands management. There have been a number of well-intentioned studies that have produced good suggestions. Unfortunately, they are based on a belief that individuals and agencies are able to work at cross-purposes to their own best interests on behalf of a larger public good or for the protection of the public landscape as a whole. This has not worked well in the past, nor is it working well as I write this. In the past, wilderness got little funding, and management of wilderness was not the way to climb the agency ladder. The consequences of neglecting the wilderness are still with us. Any incremental changes are implemented within the context of a larger system of incentives and control, and changing the priorities of how our public lands are managed becomes very difficult. Nor is the motivation to do so instilled within the agency personnel. Most are trained in some form of commodity or recreation management, not wilderness protection. Those who are trained or motivated to protect ecosystems and wildlands remain a small—albeit a growing—minority that is increasingly vocal and supported by outside groups and the public.

Implementing incremental changes will increasingly pit different groups within agencies against each other as they argue about exactly how multiple use is to be implemented; what ecosystem management is what the agency's management goal is; how responsive the agency should be to both local interests and the larger public interest; how much does the agency decide what is right, and how much does it do what it is told to do by congresswomen and senators who wield political power on behalf of various constituents and corporate groups?

These are examples of questions that will not go away if a little tinkering is done. Indeed, they may only increase. The "tinkering around the edges" solution raises hopes and expectations of self-education and internal reform within agencies that I find hard to accept. The most likely result will be disappointment, bickering, and a continuing politicization of the management agencies. This may be the easiest path for all concerned to follow. Political and agency inertia may dictate that this scenario provides an outline of what future wildlands management will look like. If so, opportunities will have been passed over or considered only superficially.

A third scenario suggests that the time is ripe for a major reassessment of how our wildlands are managed. There are a range of alternatives that might be suggested. The one that keeps rearing its head is to sell or give the federal wildlands to the states or to privatize them completely. I find the recent arguments by different shades of the "new resource economists," or "environmental Darwinists" uncompelling, simplistic, and wrong-headed.

They keep coming up, not because they are efficient solutions to wilderness management, but because they assume that only easily priced goods have value. Even within the economics fraternity, they represent a minority view. Most economists do not call for privatizing public wildlands. The privatization argument is based on outdated and untested assumptions, and often consists of very suspect or "bad" economic analysis. Except for exceptional cases, the privatization argument should be put to rest.

There remain a number of ways of improving the management of public wildlands. Some ways have been alluded to in earlier chapters, while others have not. I will focus on the major changes that could lead toward a better, and yes, more efficient management strategy. It is one person's vision, and it is not necessarily original, the most complete, or the best one. It does point toward a shift of direction. Starting on a new path can be an exciting experience, and some are already moving down it. Getting others to do so requires modifying or throwing out some cherished practices and organizational structures.

To begin with, there is a need to move beyond multiple use as a management strategy. Rather than rationalizing why it has not worked well, let's just admit that generally it has failed. Even defining what it means in concrete terms puts us on a very slippery slope. In most cases it has meant maximizing one use, most recently timber harvesting, while trying not to create too much damage for other uses. The American West is full of that kind of damage, whether to landscapes, salmon, or the habitats of spotted owls and numerous other creatures. Pretending or trying to manage for all uses under the rubric of multiple use has not worked. It was a noble, but flawed, concept, and there is no reason to spend much time trying to ascribe blame or participate in finger-pointing exercises. Agencies such as the Forest Service face severe management problems because agencies charged with multiple objective management do not work well in today's environment.

Instead of multiple use, dominant use should be the major strategy for managing our federal wildlands. It is much more simple, direct, and honest. It is also how most other agencies operate. The National Park Service, for example, has the simple task of managing the parks. The Park Service does not have to satisfy multiple use, but it does face the conflicting objective of promoting visitor use and enjoyment, while preserving the parks for future generations. As a result, it has yet to satisfy both the recreation users and the preservationists.

Dominant use management, while a better approach, still is no panacea. The Environmental Protection Agency is another example.

Its mission is clear. Protect human health by protecting the general environment. It has to set standards within which individuals, companies, and governments go about doing their business. Again, how successful it has been is debatable, as is how much authority it should have. Its budgets go up or down depending upon which political philosophy is dominant in Congress and the White House. It is able to try, under a variety of circumstances, to carry out its mission because it does not change nor does it have to accommodate local politics, as do the federal land management agencies in the West. Other federal agencies operate under similar mandates. The Department of Defense is charged with defending the country and its perceived security interests. Social Security provides some insurance so that we won't become destitute in our older years. The Post Office delivers our mail, or part of it anyway. The Federal Deposit Insurance Corporation insures the banks in which we put our money so that we are not at the mercy of the market and unscrupulous persons. The Census Bureau counts how many of us there are, and what we are like. The Federal Aviation Agency is charged with making sure the planes we fly in are safe. All of these agencies perform functions of value that we take for granted. None have to do vastly different jobs in different places as do some of the foresters I described in Chapter 2.

Even if these dominant use agencies have problems and some inefficiencies, it is foolish to insist that they would operate better in some theoretical world where people could optimize decisions as situations change. Dominant use becomes what has been called a second-best strategy.[1] The first-best strategy may be known, but be inoperable or very inefficient in the real world. I suspect that, for most of the agencies of the federal government, dominant use is the first-best strategy.

Some will no doubt respond that landscapes are too complex to be managed for a single purpose. Pursuing multiple objectives is the only plausible way to manage these wildlands. And besides, what should be the dominant strategy? Can we get broad agreement? Won't it vary from place to place, and from time to time? The response from the public seems clear enough if only politicians and land managers would listen.

Survey evidence gathered over more than thirty years has shown that people want their environment protected. These trends have increased over time as more information has become available and people have educated themselves on the value of a protected environment. The migration of people toward such places provides other evidence. Protection or preservation—call it what you like—has become a highly valued good in our society, both inside and outside cities. Post-modern or not, it has become part of what we increasingly want and expect as time moves forward. To manage our wildlands without recognizing these trends is only to postpone what is inevitable. Managing within the context of an Old West is to glance backward instead of looking forward.

Federal wildlands must be managed for the protection of watersheds, ecosystems, and biodiversity. Such a dominant use approach sets the standard for which other uses of the wilderness will be judged. Can this be done within the current framework of four major agencies that manage wildlands? Probably not. This is not a popular view for several reasons.

An often-cited advantage of having several agencies manage public wildlands is the spirit of competition that is inspired between them. Tell them what their objectives are, and they will each strive to do a better job, offer better solutions, be more innovative, and so on. This has a hollow ring to it. Agencies are not teams, each trying to win a title and striving to come out on top. They may compete for a limited budget pie, but competing for the honor of being the agency that promotes the most biodiversity seems a bit far-fetched.

Each of the land management agencies in the recent past has carved out a unique role for itself: the Forest Service with its emphasis on getting the timber out, the Bureau of Land Management providing grazing rights, the National Park Service making customer enjoyment its top priority, and Fish and Wildlife both providing or trying to save various species.

The history of competition between them is limited, except for getting management control of public lands. There is no early evidence of a competition between them in implementing ecosystem management in the best and fastest way. Certainly they are not going out of their way to provide for more and more wilderness and to respond to the larger public interest. They are being goaded to respond by individuals and various environmental constituencies. Even so, they each respond differently, and when forced to act do not follow the same guidelines. This continued fragmentation of responsibility will continue to lead to uneven and inconsistent protection of our federal wilderness.

A better approach would be to have one agency charged with ecosystem management or wilderness protection and preservation. Protection and preservation would be the main objective, and other uses would be allowed only if they were not contrary to this goal.

There have been various proposals over the years to create an expanded agency called the Department of Natural Resources. This would be a disaster. The very name suggests a continued emphasis on commodity and resource use. It would perpetuate the multiple objective philosophy within a larger bureaucracy. We need an agency whose reason for existence is the protection and preservation of environmental resources within an ecosystem framework, not a revamped Department of the Interior.

The new agency responsible for our wildlands might be a new and free-standing one, or possibly be combined with another, such as the Environmental Protection Agency. It would not demand that its managers pretend to be all things to all people. Managers of landscapes and ecosystems would form the heart of the agency. Backgrounds in ecology, wildlife, and landscape restoration would be more important than expertise in forest resources or products and range management. This would make more sense than the current pretense of holistic multiple-use forest management by forest planners who have taken two or three courses in undergraduate biology.[2] Where appropriate, agency foresters would manage tree production, and not be required to become instant experts in ecological systems management.

The alternative agency approach is not new. A Wilderness Agency was proposed as part of the negotiations leading up to the 1964 Wilderness Act, but did not survive.[3] Without an agency base, much wilderness management was done almost on an ad hoc basis. If an agency had been established, it is hard to tell how successful it would have been.

Without an agency, there was no separate budget for wilderness management. With an agency, there would have been a budget, and a corps of dedicated people working together rather than as individuals in agencies where they have been at best tolerated rather than applauded for their efforts. An agency that had wilderness lands that were contained within numerous other agency lands would have been perceived as a threat to the commodity-driven multiple-use agencies.

There would have been pressures to prevent the environmental damages from the other agencies' lands from spreading into the wilderness areas. Demands for buffer zones would have built up as would have pressures to expand various wilderness area boundaries by recognizing that present boundaries may cut intact ecosystems into pieces, with extensive logging permitted in one part and none in the other. There have been far too many occasions when, while hiking in wilderness, I could tell where it ended by simply seeing where logging operations extended up to its borders and stopped.

A Wilderness Agency may have noticed early on the futility of trying to manage wilderness as unconnected patches or islands in a larger landscape, and may have began lobbying for a means of connecting them, or creating some form of an ecologically rational wilderness landscape. The issue of considering the landscape as a whole would not seem as radical today as it still does to most public wildlands managers. Citizens would have been encouraged to think of our Western public lands as a series of regions, rather than individual national forests and parks. Instead, wilderness issues were marginalized, and contained within the more important business of providing some form of resource services within each agency structure.

From the perspective of individual agencies, they were "smart" to see that a new upstart Wilderness Agency was not started during the socially active era of the 1960s. Who knows what kinds of additional problems that might have created for them. I have spelled out a few ways in which such an agency might have brought some more enlightened discussion and recognition of the need to take a more geo-ecological approach to wildlands management.

The agencies could have been proactive and leaders in how our public lands are managed rather than perceived as they are today as doing a poor job, not trusted by the public, and seen as either willing or unwilling "servants" of companies that extract commodities as a form of corporate welfare and of modern-day ranchers whose use of public lands has degraded much of it.

Given the number of devoted and hard-working people in these agencies, this characterization may be unfair, but it is a public perception and one rooted in the current conditions of our Western wildlands. Or perhaps a Wilderness Agency would have simply led to more interagency bickering instead of the interagency cooperation deemed necessary to make the system work. Clearly, it is not working very well right now.

The need to go beyond managing on a simple national forest or even agency-by-agency basis is pretty obvious to almost everyone who takes a look at the situation and is not beholden to some political entity. That is why interagency cooperation remains the linchpin of almost all recommendations made. Calling for the elimination or merger of functions into a different agency is hard for many to do because of the past corruption and stigma tied to the Department of the Interior.[4]

James Watt, the controversial Interior Secretary during the Reagan Administration, resigned, later was indicted, and pleaded guilty in 1996 to a misdemeanor charge in return for some 25 other charges being dropped.[5] The history of the department is tainted, and will continue to be so, as long as its primary purpose is to provide production subsidies and control extractive activities on the public lands in its jurisdiction. The Forest Service, which is in the Department of Agriculture, has its own set of problems. The Department of Agriculture also has a history, and is in some ways comparable to the Interior Department, with its own subsidy structures and an interesting variety of scandals in its background. Oversight and management has to be given to an entirely new agency.

Another reason why some people get anxious about considering a new wildlands management agency is the fear that politicians will take advantage of any restructuring and downsizing to eliminate whatever protections these agencies currently provide. After all, agencies have changed in the past and can be reformed to meet the demand of a new public will. The privatization crowd will jump in and, if not give away, sell off much of the public wildlands at fire sale prices. If not that, any new agency will be set up in a very weakened form, and will be charged with operating within a "people come first" multiple-use agenda. Management oversight, and the ability of outside groups to challenge decisions will be weakened. Better to stay with familiar agency structures than risk striking out into unfamiliar territory.

These are valid concerns, and that is why there should be a larger public discussion and debate on exactly how people want their public lands managed. Despite whatever political changes have taken place, when politicians ask their constituents if they want environmental laws weakened, the answer is a resounding no. During the 1980s and 1990s corporate lobbyists have tried with very mixed success to rewrite the environmental laws and regulations under which they have to live. In a democratic society we wouldn't expect them to do otherwise, even if their money has a corrosive effect on our quality way of life. Similarly, the individuals and corporations who benefit from extracting resources from public wildlands can be expected to try to influence the legislation under which they can get windfall gains from public lands. And for them, the ride on the public lands gravy train has been a long one. The laws under which private profits can be made on public lands are outdated and an overhaul is long overdue.

Charles Wilkinson calls these antiquated laws "the lords of yesterday."[6] They are the product of the mid- or late-nineteenth century when legal solutions of the time reflected the attitudes that nature was to be conquered and natural resources were commodities. Today, attitudes have changed dramatically but the laws embodying the old beliefs remain. Conflicts over the consequences of these outmoded laws have led to pitched battles between environmentally oriented migrants, residents, and Westerners who are trying to preserve economic interests that were sanctioned by laws and social mores of the past.

These lords of yesterday include the Mining Law of 1872, which allows individuals and mining corporations to go onto federal wildlands and extract hard rock minerals, including gold, silver, uranium, copper, and many others, for free. There is no charge. They can even get title to the land over a deposit if they discover valuable minerals. Other lords include laws and regulations that promote the overgrazing of public lands; subsidize massive road building into wildlands, promote the over-cutting of forests, allow the building and subsidizing of dams that destroy fish runs, and perpetuate outdated water laws.

The water lord of yesterday is particularly contentious. The people who got there first in effect own the property rights to the water, and these cannot be taken away without the government paying them for it. The American West depends on water, but unlike the watery East where the riparian water doctrine requires sharing water between users, they need not do so in the West. This has created any number of conflicts and controversies in the West that have filled volumes of books. The federal government has helped out by subsidizing projects that contribute to the misallocation of much of this water.

There have been efforts to change the outrageous continuation of benefits at little or no cost based on conditions associated with the settling of the West. To date, the will of the majority as expressed in public opinion polls has not prevailed. Western politicians have been able to block these changes, but it is only a matter of time before the changes are implemented and the laws revised. These laws are obscure and unfamiliar to most Americans and the agencies managing our wildlands have not been forthright about revealing the implications of these laws as they relate to our Western landscape.

Congress would be well advised to modify these laws (the sooner the better) to take itself out of the management picture. In most cases, Congress passes the laws and allows the agencies to manage according to the dictates of the laws. If it does not like the results, it can go back and modify the law. In the meantime, normally it does not tell the agencies how they should manage, or what standards should be passed.[7] Unfortunately, politicians interfere with regularity when it comes to setting wildlands management policy. There are any number of cases in which Western politicians have not liked how many million board feet of timber were going to be cut in a forest plan, have intervened, asked for another look, got it, and subsequently the number of trees cut increased substantially. In the case of the Tongass National Forest in Alaska, they dictate what the cut will be each year.

Responsible members of Congress have to realize that this has to stop. How can there be any talk about science dictating what can and cannot be done in a particular place when it can be overridden by politics that are based on either no science, or a questionable "gray science." Congress does not allow its members to go around telling the Environmental Protection Agency what the air pollution standards should be for their specific city, so why should members of Congress be allowed to set policy for each national forest or grazing district that they have an interest in? They shouldn't. It only feeds the cynical conviction of citizens, the agencies, academics, nongovernmental organizations, and others that only politics really matters. Politicians should get out of day-to-day management.

Congress also has to move beyond its traditional approach of treating the decisions about these wildlands as though they stopped at the border, at which point Congress must defer to state congresswomen and senators. Citizens of these states often act as if they have more *rights* to these lands simply because they live near them or use them. In their opinion, they have more of a right to decide how the lands should be used than citizens who live far away. Clearly, they have no such right. If anything, they have had the *privilege* of gaining the benefits from these lands either by using them at almost zero cost for recreational or other purposes, or, in the case of a select few, by using them to make a living and maintain a lifestyle of their own choosing.

In more and more cases their property values also are increasing. The amenity and cultural lifestyle values of living in proximity to these large public wildlands simply are not available to most of the population. These are unpriced subsidies, and it would be hard to allocate an exact price to some of them. Those of us who value this lifestyle highly count ourselves lucky, and should thank others who allow us to benefit from living surrounded by wilderness. We should not, however, assume that we can tell the rest of the country that we have the *right* to tell them how the nation's wilderness should be managed. There is a difference between a *right* and a *privilege*, and some Westerners have to be more humble in acknowledging that difference.

What Americans, whether they are Westerners or not, are saying is "protect our wildlands." Doing so requires moving toward the large landscape ecosystem approach suggested by the Wildlands Project and the "newer voices" within the agencies themselves. Casting politics aside, how should we protect watersheds, what kind of regional management strategies are needed irrespective of where agency and private land boundaries start and stop? These are among the exciting new questions that a new agency needs to ponder, and that the scientific community needs to debate in journals and books that set new standards up on which decisions will be based. A more map- and landscape-oriented approach should dictate policies, not how many jobs can be provided.

At the same time, simple decency, equity, and a sense of responsibility dictate that the needs of localities that are dependent upon the subsidies of the Old West need to be considered. Local people need to be involved from the start in a long-term process to evaluate how protecting ecosystems and landscapes will benefit them. They need to be convinced that the vitality of their local communities depends on making the transition from a commodity-based strategy to a protective strategy.

Local people should share the responsibility of protecting the wilderness around which they live since they are the ones who primarily reap the benefits. In the past many people were simply "hired hands" who worked for local or non-local companies extracting resources and had little responsibility for contributing to the protection of federal wildlands. They should contribute to the monitoring of conditions on lands with which they are familiar. To carry out such responsibilities, training and educational programs will need to be instituted early on in the process.

I am convinced from talking to Westerners that most of these people live in the more rural interior West because of the physical environment and a unique way of life that would be difficult to duplicate elsewhere. They want to work in or around the woods, ranch, or hamlet. They don't want to work in fast-food joints or motels, but there are plenty of other jobs that will be associated with an outdoor life in monitoring ecosystems and practicing gentler and smaller-scale logging and ranching operations.

The West is changing. Its local communities are or will be changing, and those who love living where they do will adjust. They don't want to move on. Living in the inner West is more important than being a logger in Georgia or a lawyer in Chicago, Atlanta, or New York. Somehow they will have to stay and confront the changing New West and the people it attracts.

They are getting a very small taste of what it must have been like for the original settlers of these lands to watch the white man come, first washing over and then displacing them. The current "old Westerners" will not be displaced, but their communities will change. They have a choice the Indians did not have. Will they contribute to that change and help guide it, or will they symbolically wail and fight imaginary Wars on the West?

Their chances of winning or stemming the tide are better than those the Indians had. They at least have some political representation, and often it is powerful. I don't think they can or should try to fight the changes taking place in the West. They also should look toward their Native American neighbors, from whom they have largely been estranged, and see if they can find better ways to share the "wild West." Ignoring the role of Native Americans will lead to more strife, and this time the white man and his allies will be on the defensive.

## NATIVE AMERICANS AND WILDERNESS

Native Americans have been ignored for too long in the management of the lands that were formerly theirs. They have begun more aggressively to assert their treaty rights, and also to sue to protect the natural environment upon which their societies used to depend. Most of the development strategies tried by the tribes have not built on this cultural history.

The current development trend among Indian tribes that centers on gaming facilities carries the promise of economic gain, but perhaps at the price of great cultural strife. Recognizing these dangers, the Navaho tribe—over a quarter of a million strong—has resisted the urge to make gaming their main source of revenue, with tribal members voting against a measure to have gambling on the reservation. A few tribes are making large amounts of money and providing jobs for Indians. The money pays for better schools, housing, and a variety of social services, but at the price of increased corruption among Indians themselves. It would be foolish to expect otherwise. Taking money largely from white men and women may be satisfying and a fair turnabout, but it does bring its own social and cultural risks. However, the choice should be theirs and not imposed on them by a state or even a paternalistic federal government.

It is not idealistic to suggest that pursuing training and employment in ways more in tune with the tribal past might bring greater long-term benefits to the tribes. Given the pervasiveness of the dominant materialistic culture, it is amazing that more Indians have not given themselves over to pursuing the American dream of accumulation of wealth and job status. Even as native languages are under siege and the continual threat of extinction, the traditional cultures are carried on in the powwows where young and old perform the dances that have been passed down through the generations. Whenever I attend a powwow, I find it packed with people of all ages, not just the older people who are trying to keep the traditions alive.

The presence of people, young or old, on reservations who are trying to uphold some form of a more traditional way of life, and of some tribes practicing what we now call sustainable forestry, is evidence that our public wildlands could have been better managed if more cooperative rather than paternalistic efforts had been made on the part of the non-Indian managers and their teachers to learn from the tribes. It is not too late to be humble and say "teach us, or let us learn together." Unfortunately, as has been demonstrated time after time, this has not been part of the program of federal land managers.

Wilderness management has not worked very well under current schemes. Why not do a set of "experiments" that either incorporates Native Americans into cooperative wildlands management schemes, or allows them to have full management responsibility. In many areas the federal lands are in close proximity to tribal lands. Cooperative or joint management of these lands would provide a more integrated policy than the segregation of land-based ownership patterns. After all, the land used to be theirs. A new approach just might build a more trusting relationship between Indian and non-Indian managers than currently exists.

A shift toward involving Indians more directly in how our public wildlands are managed would benefit everyone. It would provide an emphasis as well as an outlet for the more traditional factions of the tribes to show that playing the white man's game of producing commodities is not the only game in town. Harvesting trees, mining minerals, or providing gambling activities are not the only ways of promoting self-development.

These would be experiments ranging from a simple inclusion of Indian lands in a larger ecosystem approach, to working with them to manage currently owned federal lands in the traditional ways as they see fit. The landscape of the Indians or First Nations was different from the federal wildlands of today.

We can learn from the past practices of the Indians, who used fire deliberately to clear land, provide habitat for hunting, promote growth of various plants and restrict others, among other things. The first European settlers and later academics derided the Indians for burning the landscape.[8] In earlier times, and still today, people have often spoken as if fires were everywhere and the burning was homogenous force spreading out over the land. This was an oversimplification of a complex burning process that varied by tribe, season, location, and purpose. Today's scientists, using more controlled experiments, are trying to copy what tribes did as part of their everyday lives. Excluding the Indians—again—seems a bit unfair.

The practice of fire suppression by the Forest Service and other agencies is now considered to have been disastrous and one example of the failure of natural resource science. Today, natural resource scientists are claiming that one result of those forest suppression policies has been to turn many forested wildlands into tinderboxes that are ready to explode into huge firestorms unless we trust forest scientists and engineers with their salvage logging and prescribed fires to reduce the risk. Whether much of this is hyperbole or not, it does in retrospect show that the Indian way was more natural and better for the landscape in the long run. The tribes did not have the benefit of European science, but they had spent thousands of years learning about and adapting to their environment.

The suppression of Indians and fire went hand in hand. Surely, moving toward a more enlightened policy should involve joint management and learning between agency managers and Indians. Fire policy and protection of habitats are natural places to begin.

Another way in which both Indian reservation lands and federal wildlands are important to Indians is for spiritual purposes.[9] Indian religions make much use of spiritual places, and carry out initiation rituals over longer periods of time (months) than the Judeo-Christian religion of the dominant culture, for which there are few spiritual places in the United States.[10] For many Indians their religious practices are tied to sacred places and vision quest sites, that are situated in forests, mountains, or deserts. There are specific places that carry a religious significance, and often these are not on the reservation but on federal or private lands. Obviously, this can create a great problem for Indians who are trying to participate in spiritual practices passed down over generations, especially when these lands are mined, logged, or taken over by the advancing forces of urbanization or streams used for religious cleansing are silted.

In terms of religion, Indian reservations generally are divided between those who accepted the faith of the missionaries and those who have maintained the traditional ways. Many whites and some land managers simply do not take the traditional religion seriously. Sacred mountains and the body of Mother Earth mean little to them, as do places like the Heart of the Monster, the equivalent of the Garden of Eden for the Nez Perce, the place where it all began. Land managers do not want to hear about sacred sites if it disrupts a mining or logging operation. As one Indian poignantly said,

> Imagine if you had to go to court, explain and defend to an alien people your Bible and Christianity in a less than an hour, and then some judge would decide whether your religious practices were valid. That is what I have just had to do.[11]

With a few exceptions, Indian efforts to get greater recognition of their religious rights have not met with much success. The participation of tribes in the management of lands that are important to them and consideration of places of spiritual value would be much easier if the focus was on an ecosystem approach. Previously Indians were part of that larger ecosystem. They were a part of their natural surroundings.

The spiritual question puts Indians in a Catch-22 situation. Much depends on these sites and rituals being kept secret from outsiders. If they put all of their places that have spiritual value onto a map, their traditions and practices are diminished. That is why they ask for recognition of their cultural resources without their having to reveal all their places and associated practices.

Only when religious sites are threatened by commodity extraction, development, or tourist use do the Native Americans raise legal issues and try to protect them. In changing the way we manage the lands, we need also to change the way in which we have viewed the native peoples and to make them an important part of how we go on using the lands.

## WILDERNESS, NATURE, AND SOCIETY

What we consider wilderness today is much different than what it used to be under Indian occupancy. Our notions of wilderness as pristine and pure, and of the need to keep it that way, may be noble, though somewhat unrealistic. If we had not forced the Indians off much of the land that we now consider wilderness, it would look different than it does today. Does that make it less wild? Probably not.

Efforts to make the wildlands more wild and natural raise a host of vexing questions. What state do you choose, and how much deviation from it do you allow as you try artificially to bring it back? What do you do about introduced plants? Do you try to bring animals back? Efforts to bring wolves and grizzly bears back are based on notions that they belong there.

Wolves were exterminated, killed by the thousands in unnatural ways. The land managers decided they were a nuisance, a threat to be eliminated, and set up bounties and other incentives to get rid of them. Now, years later, efforts are made to bring some packs back, campaigns are mounted, and wolf t-shirts are sold. Ranchers, who were among those who originally advocated that the wolves be exterminated, are indignant. Protect us, not wolves. But times have changed. Wolves make for better pictures on t-shirts and are more vital to the ecosystem than the rancher, especially a polluting one. Wolves also are wilder.

Wolves were eliminated in part because they had no value for ranchers and farmers, or for hunters. Whereas deer, elk, moose, buffalo, and bears have value to hunters and outfitters, the wolf was regarded as a competitor that was hunting for many of the same species as the human hunter. After the settlement of the West, states set up fish and game departments to protect and maintain enough wildlife, from grouse to bears, to satisfy the requirements of hunters. The value the animals have is a commercial one, and Western states make sure their wildlife program is properly funded.[12]

Without natural predators such as wolves that cull out the weaker individual of the species, the herd populations increase dramatically, increasing the risk of decimation by disease, as has happened in Yellowstone National Park. In these instances, special hunting permits are issued or a short hunting season is opened to reduce the herds. Naturally, hunters prefer to serve as the means of population control rather than allowing natural predators to do so.

A wilderness without bears or wolves is less wild, no doubt about it. How much wildness do we want? Creating tree farms on public lands diminishes wildness, and makes these lands more like the industrial society upon which we depend. Wildness and the American West provides a counterpoint to our highly urbanized civilization. Paradoxically, we established wilderness because of that civilization. It is not the original wilderness or the Indian wilderness. It is the wilderness we have tried to freeze at a particular point in time. It is a place we can go to, and a very important one in our culture.

Wilderness and our wildlands are an important part of our landscape. I argue that in the American West it is a critical part if we still want to maintain the West as a viable region. Urban growth and expansion in the West, as in other regions, of the country cannot be stopped. Instead it is surrounded by wildlands and wild spaces. Recognizing this and protecting and incorporating these wildlands into a regional landscape will allow the West to maintain its unique identity.

Wilderness is importance to a variety of people, especially within the context of the social and economic changes taking place in the region—whether a hiker is getting physical exercise, spiritual enlightenment, or a simple awareness that she is in a place where she does not reign. Wild nature will go on whether someone is there or not. She doesn't need us. Plato's question of whether a tree falling in the forest makes a sound if no one is there to hear it seems a silly one when you are considering wild nature. The fact that she has no need of the human being gives these wilder places their spiritual qualities. Civilization and industrialization can stand outside.

All kinds of people seek some form of escape into or near wilderness, not just hikers and backpackers, fishermen, miners, or loggers. It has become a form of escape for neo-Nazis, constitutional patriots, religious zealots, and others seeking an escape from a mainstream society that alienates them. Some move to be nearer to "whiteness."

They bring with them false myths about searching out the new American frontier. They are searching for the last good places where men can be men, and government is not on your back. They have images of being independent, throwing away their urban skins, and building newer forms of local democratic societies where people will think as they do. Democratic for them, less so for others. Or they will simply hide out, and fight back if the larger world infringes or closes in upon them.

The projecting of false myths and stereotypes into the present can be dangerous. Forgetting about the persons displaced in the past in the name of settling the frontier, and the exploitation of minority and immigrant groups is not a past we want to see repeated. The shameful treatment and displacement of Native Americans as well as Hispanic, Chinese, and Japanese peoples, is not a past we should ignore as groups move into today's West and preach about White Nations, the discriminated against white male, and the need to fight back. This is a perversion of what the New West is becoming.

What all the different groups moving into the West have in common and what they will have to share is the use of the public wildlands that will continue to define their lives. Some people may simply look at it out of their house windows, or as they drive through the different spaces of the West, while others will recreate in it or seek solace from it for the deeper meanings of their lives. Whatever and however they interact with it, they will not be able to ignore the physical features that define the West. Hopefully, they will get to know each other and work together to build places and lives that are fulfilling.

The physical landscape of the West offers great hope for developing places that people believe they can belong to. There are plenty of people in the region who already feel that way. The wildness in the public lands provides a basis for life that too many people take for granted. Protecting that wildness rather than subduing it would be a unique way of promoting the vitality of both communities and cities throughout the region. Evidence shows that the region will continue to grow. That very growth would threaten the natural environment that defines the region were it not for the federal wildlands that serve as the core feature attracting people into the region.

Imagine that the privatization argument held sway and the public lands were sold off to the highest bidder. Would the American West look much different? Would environmental groups raise money and outbid corporations for much of the land? Would it remain a land of wilderness and less development than under federal management? Some economists claim that, if the subsidies were eliminated, most of the public lands have little value for timber harvesting because the costs exceed the revenue to be made.

It is hard to know exactly how the West would be affected. Certainly by the criteria of other places it is not overpopulated. Population densities are very low and there is plenty of open space. In the non-arid West, much of the federal estate could be profitably subdivided into ranchettes or second homes, or rustic cabins for urbanites wearing designer flannels and snake skin cowboy boots. One estimate is that the Northwest region could easily accommodate an additional 25 million people.[13] But do we want that many people in the region? Does it remain the same even if that many people were to move in? The more important question is do we want the market to determine what the landscape of the West looks like?

Do we let the price of timber and minerals, cows, and houses determine how much of the American West is cut down, dug up, overgrazed, and built upon? In such a West only commodities matter. That is a developmental logic where the "unseen hand" rules, and not a public citizenry. Or do we use other non-economic criteria to decide how we want to use the public wildlands? Will we place a higher priority on using the public wilderness to establish ecological values and limits? Will we use our crude ecological and ecosystem management science to establish geographical limits to which we will conform? Will we err on the side of wilderness protection and preservation, allowing nature more latitude in dictating what its future will look like?

The past predisposes us to think we can manage or engineer our wilderness. We have a faith in our ability to control environmental and social costs. We can always call in the mega-engineers to manage deserts, forests, or waterways. Much of the urban West exists because of their work. Nature is slowly striking back, and the odds and history are on her side.[14]

The public wilderness provides an opportunity to set an example of how people and landscapes can evolve without having to pit one against the other. The United States has given the world the concept of national parks and wilderness. Indeed it is one of our noblest achievements. Today, how we view and treat the wild mirrors our ability to provide the basis for a more humane society. Thoreau's "in wildness is the preservation of the world" still provides us a maxim that we can ignore only at our own peril.

# Notes

## CHAPTER ONE: Wilderness and the American West


1. Alfred Runte, *National Parks: The American Experience* (Lincoln: University of Nebraska Press, 1979).
2. W. Nugent, "Where Is the American West? Report on a Survey," *Montana, The Magazine of Western History*, Summer 1992, 2–23.
3. Sarah Deutsch, "Landscapes of Enclaves: Race Relations in the West, 1865–1990," in William Cronon, George Miles, and Jay Gitlin, *Under An Open Sky, Rethinking America's Western Past* (New York: W.W. Norton, 1992): 110–131.
4. Donald W. Meining, "Spokane and the Inland Empire: Historical Geographical Systems and A Sense of Place," in David Stratton, ed., *Spokane and the Inland Empire: An Interior Pacific Northwest Anthology* (Pullman, WA: Washington State University Press, 1991).
5. Mike Davis, *City of Quartz: Excavating the Future in Los Angeles* (New York: Vintage Books, 1992); Edward Hoagland, *Balancing Acts* (New York: Simon and Schuster, 1992). For a post-modern view of Los Angeles, see Edward Soja, *Postmodern Geographies* (London: Verso, 1989).
6. Christiane von Reichert and Gundars Rudzitis, 1992, "Multinomial Logistic Models Explaining Income Changes of Migrants to High Amenity Counties," *The Review of Regional Studies*, vol. 22 (1992): 25–42; Gundars Rudzitis and Rosemary A. Streatfeild, "The Importance of Amenities and Attitudes: A Washington Example," *Journal of Environmental Studies*, vol. 22, 1993. 269–277.
7. Frederick Jackson Turner, *The Frontier in American History*, Wilbur Jacobs, editor (Tucson: University of Arizona Press, 1986).
8. Dayton Duncan, *Miles From Nowhere: In Search of the American Frontier* (New York: Penguin Books, 1993).

9. For a good overview, see Michael P. Malone and Richard W. Etulain, *The American West, A Twentieth Century History*, Lincoln: University of Nebraska Press, 1989).
10. Michael E. McGerr, "Is There a Twentieth-Century West?" in William Cronon, George Miles, and Jay Gitlin, *Under An Open Sky: Rethinking America's Western Past* (New York: W.W. Norton, 1993).
11. For an intriguing description of this process, see the two articles in the Spring, 1992 issue of *Montana: The Magazine of Western History*—Gerald Nash, "The West in Historical Perspective," pp. 3–16; and Julian Crandall Hollick, "The American West in the European Imagination," pp. 17-20.
12. Personal communication, Christiane von Reichert.
13. Karen Kaasick Dean, *The American West: A View From Estonia*, Master's thesis, Department of Geography, University of Idaho, 1995.
14. Patricia Nelson Limerick, *The Legacy of Conquest: The Unbroken Past of the American West* (New York, W.W. Norton & Company, 1987). Limerick was not alone in arguing for a new interpretation for western history. Among other notable historians who have been working in a similar vein are Donald Worster and Richard White, who in a recent book refused even to use the term "frontier," or the "f word" as it has become known among these and other new Western historians. There was, of course, the typical reaction from other historians who said that what these new historians of the West were writing about was not new. Others had done it all along, though perhaps not to the same extent. Perhaps, but the new historians placed a wider range of groups at center stage in their view of western history.
15. Susan Armitage, 1995, "The Old/New Drama of Western History," *Universe*, Fall, Vol. 7, 18–19.
16. Gundars Rudzitis, 1991, "Migration, Sense of Place, and Nonmetropolitan Vitality," *Urban Geography*, vol. 12 (1991): 80–88; David Mura, "Strangers in the Village," in Rick Simonson and Scott Walker, eds., *Multi-Cultural Literacy: Opening the American Mind* (Saint Paul: Graywolf Press, 1988), pp. 135–154; and in same volume, Ishmael Reed, "Multi-Cultural Literacy: Opening the American Mind," pp. 155–160.
17. Susan Armitage, "Through Woman's Eyes: A New View of the West," in Susan Armitage and Elizabeth Jameson, eds., *The Woman's West* (Norman: University of Oklahoma Press, 1987), pp. 9–18; Jeanne Kay, "Landscapes of Women and Men: Rethinking the Regional Historical Geography of the United States and Canada," *Journal of Historical Geography*, vol. 17 (1991): 435–452.
18. Gundars Rudzitis, "Migration, Sense of Place and the American West," *Urban Geography*, vol. 14 (1993): 574–584.

19. See Susan Armitage note 16.
20. See Gundars Rudzitis, notes 16 and 18; Jack Davis, "Civilizing the Europeans," *Idaho*, vol. 9 (1992): 13; William Irwin Thompson, *Imaginary Landscape: Making Worlds of Myth and Science* (New York: St. Martins Press 1989); Pierce Lewis, "America Between the Wars: The Engineering of a New Geography," in Robert D. Mitchell and Paul Groves, eds., *North America: The Historical Geography of a Changing Continent* (Totowa, N.J.: Rowman and Littlefield, 1987), pp. 410–428; Charles F. Wilkinson, "Law and the America West: The Search for an Ethic of Place," *Colorado Law Review*, vol. 59 (1988): 401–426.
21. The classic study on attitudes toward wilderness was Roderick Nash, *Wilderness and the American Mind*, New Haven: Yale University Press, 1967 and subsequent editions; A more recent book also destined to become a classic is Max Oelschlaeger, *The Idea of Wilderness: From Prehistory to the Age of Ecology*, New Haven: Yale University Press, 1991. He disagrees with Nash on several key points and I recommend both books and their extensive bibliographies to anyone. The more adventuresome should consider as well Hans Peter Duerr, *Dreamtime: Concerning the Boundary between Wilderness and Civilization*, translated by Felicitas Goodman, Oxford: Basil Blackwell, 1985; For those wanting to venture back to the Greeks and their attitudes the classic work is Clarence J. Glacken, *Traces on the Rhodian Shore: Nature and Culture in Western Thought from Ancient Times to the End of the Eighteenth Century*, Berkeley: University of California Press, 1967.
22. For examples on the importance of nature in urban as well as rural areas, see Tony Hiss, *The Experience of Place* (New York: Vintage Books, 1991); Barry Lopez, "The American Geographies," in Robert Atwan and Valeri Vinokurov, eds., *Openings: Original Essays by Contemporary Soviet and American Writers* (Seattle: University of Washington Press, 1990), pp. 55–70; Mojdeh Baratloo and Clifton J. Balch, *ANGST: Cartography* (New York, Lumen Books, 1989); Yi-Fu Tuan, "Place and Culture: Analeptic for Individuality and the World's Indifference," in Wayne Franklin and Michael Steiner, eds., *Mapping American Culture* (Iowa City: University of Iowa Press, 1992); Thomas Michael Pyle, "Intimate Relations and the Extinction of Experience," *Left Bank*, vol. 2 (1992): 61–69.
23. George Perkins Marsh, *Man and Nature: Or Physical Geography As Modified By Human Action* (New York: Scribners, 1864). Marsh's analysis is particularly appropriate for the 1990s as Eastern Europe and the former Soviet Union provide an updated version of the scary consequences of unhindered state-directed industrial growth that is carried out under the banner of economic growth.

## CHAPTER TWO: History and Management of Wilderness

1 There are a number of good historians of forests in America. A good place to start is M. Williams, *Americans and Their Forests: A Historical Geography* (Cambridge, MA: Cambridge University Press, 1989).
2. M. Frome, *Battle for the Wilderness* (New York: Prager, 1974); C.W. Allin, *The Politics of Wilderness Preservation* (Westport, CT: Greenwood Press, 1982).
3. G. Rudzitis, "Federal Lands: Wilderness Management Policy," *Environment*, vol. 29 (1984): 2–4.
4. G. Rudzitis, "Contention or Cure? Is Public Ownership A Panacea for Protecting the Maine Woods?", *Habitat*, 1990, pp. 28–31.
5. M. Frome, "The Conditions of Wilderness," in M. Frome, ed., *Issues in Wilderness Management* (Boulder: Westview Press 1985): 1–6.
6. Gifford Pinchot was considered quite liberal for his time, and whether he would wear the same liberal mantle today would be speculation. He attacked corporations and their control, but in a famous "battle" with John Muir fought for "progress" and putting a dam in a sister valley to Yosemite to supply water to San Francisco. Pinchot won and the split between preservationists who look back to John Muir and conservationists who trace their lineage to Gifford Pinchot continues today.
7. A review of the *Journal of Forestry* since 1980 shows that some still feel that way. In their minds, what the forestry profession needs is better public relations, to explain to the public why what they are doing is right. The public rather than the agency needs to adapt. Others suggest that public attitudes have changed and that the agency needs to adapt to these new attitudes.
8. S.P. Hays, *Conservation and the Gospel of Efficiency: The Progressive Conservation Movement 1890–1920* (Cambridge: Harvard University Press, 1959).
9. In that regard Gifford Pinchot was a rare individual, and an exception to the rule of the generally inconspicuous heads of the USFS who remained generally unknown to the general public, and most historians as well.
10. Paul W. Hirt, *A Conspiracy of Optimism: Mismanagement of the National Forests Since World War Two* (Lincoln: University of Nebraska Press, 1994).
11. C.O. Sauer, "The Agency of Man," in W.L. Thomas, Jr., ed., *Man's Role in Changing the Face of the Earth* (Chicago: University of Chicago Press, 1964), 49–69.
12. T.M. Power, *The Economic Battle for the Landscape of the American West* (Washington, D.C.: Island Press, 1996).
13. P. Shabecoff, *A Fierce Green Fire: The American Environmental Movement* (New York: Hill and Wang, 1993).

14. D.A. Clary, *Timber and the Forest Service* (Lawrence: University of Kansas Press, 1986); D.H. Henning, "Wilderness Politics: Public Participation and Values," *Environmental Management*, vol. 11 (1987): 283–293.
15. A. Leopold, *Sand County Almanac* (New York: Oxford University Press, 1966). For the views of a modern day Thoreau and Leopold rolled into one see Gary Snyder, *The Practice of the Wild*, San Francisco, North Point Press, 1990.
16. See Rudzitis, note 3.
17. C.F. Wilkinson, "Law and the American West: The Search for an Ethic of Place," *University of Colorado Law Review*, Vol. 39, pp. 401–425. For examples and discussion of the inability of the agencies to cooperate, see the papers in David Lime, ed., *Managing America's Enduring Wilderness Resource* (St. Paul: University of Minnesota Extension Service, 1990).
18. The extent to which such issues remain politically charged is shown by a pledge by President Clinton to not allow mining outside of Yellowstone after he vacationed in the park during the summer of 1995.

## CHAPTER THREE: Ecosystem Management and Beyond

1. Hal Salwasser, "From New Perspectives to Ecosystem Management: Response to Frissell et al. and Lawrence and Murphy, " *Conservation Biology*, vol. 6, no. 3 (1992). I am indebted to John Hintz for his research assistance on the attitudes toward ecosystem management discussed in the first part of this chapter. As reported by historian Paul Hirt, 1989 was the year Chief Robertson began emphasizing a commitment to a new direction and vowing to be more environmentally responsible. See Hirt's 1994 book, *A Conspiracy of Optimism: Management of the National Forests Since World War II* (Lincoln: University of Nebraska Press).
2. Jeff DeBonis, "Ecosystem Management Needs Clarification," *Environment*, December 1994.
3. Harry V. Wiant, "Ecosystem Management: Retreat from Reality," *Forest Farmer*, vol. 54 (1995).
4. William C. Siegal, "America's Forests: A Vital Economic Resource," *Journal of Forestry*, February 1995.
5. Allan McQuillan, "National Public Tree Farms: Towards a Spectrum of Designations," *Journal of Forestry*, January 1994.
6. Dean S. DeBell and Robert O. Curtis, "Silviculture and New Forestry in the Pacific Northwest," *Journal of Forestry*, December 1993.
7. Lloyd C. Ireland, "Getting from Here to There: Implementing Ecosystem Management on the Ground," *Journal of Forestry*, August 1994.
8. J.H. Cissel, F.J. Swanson, W.A. McKee, and A.L. Burditt, "Using the Past

to Plan the Future in the Pacific Northwest," *Journal of Forestry*, August 1994.

9. See, for example, John Zasado, "Understanding Ecosystem Management," *Journal of Forestry*, July 1994; V. Alaric Sample, "Why Ecosystem Management?" *American Forests*, July–August 1994; Christopher Wood, "Ecosystem Management: Achieving the New Land Ethic," *Renewable Natural Resources Journal*, Spring 1994.
10. Anna Marie Gillis, "The New Forestry," *Bioscience*, vol. 40 (1990); Tim Foss, "Alternative Forestry," in Mitch Friedman and Paul Lindholt, eds., *Cacadia Wild: Protecting an International Ecosystem* (Frontier Publishing, 1993).
11. Thomas R. Stanley, Jr., "Ecosystem Management and the Arrogance of Humanism," *Conservation Biology*, vol. 9 (1995).
12. R. Edward Grumbine, "What is Ecosystem Management?" *Conservation Biology*, vol. 8 (1994); Amy Kerr, "Ecosystem Management Must Include the Most Human of Factors," *Bioscience*, vol. 45 (1995).
13. John Hintz, "Ecosystem Management and the Future of Our National Forest," unpublished paper.
14. There are a number of articles describing Gap analysis. A good place to start is J.M. Scott, B. Csuti, J.D. Jacobi, and J.E. Estes, "Specie Richness: A Geographical Approach to Protecting Future Biological Diversity," *Bioscience*, vol. 37 (1987): 782–788; and J.M. Scott, B. Csuti, K. Smith, J.E. Estes, and S. Caicco, "GAP Analysis of Species Richness and Vegetation Cover: An Integrated Biodiversity Conservation Strategy," in K.A. Kohm, ed., *Balancing on the Brink of Extinction: The Endangered Species Act and Lessons for the Future* (Washington, D.C.: Island Press, 1991), 282–297.
15. For example, see Henry L. Short and Jay B. Hestbeck, "National Biotic Resource Inventories and GAP Analysis: Problems of Scale and Unproven Assumptions Limit a National Program," *Bioscience*, Vol. 45 (1995): 535– 539.
16. There is also a GAP manual that spells out in some detail what is involved in doing the analysis. See J. Michael Scott, ed., *A Handbook for GAP Analysis*, Idaho Fish and Wildlife Research Unit, University of Idaho, 1994.
17. When at a conference in 1983 I asked Max Peterson, then Chief of the Forest Service, about the use of buffer zones for wilderness. He replied that he felt they were unnecessary, and the USFS could protect wilderness quite well. There has been no significant change in the USFS policy since then.
18. Mark W. Brunson, "The Changing Role of Wilderness in Ecosystem Management," *International Journal of Wilderness*, vol. 1 (1995), 12–15, quoting H. Salwasser, *Journal of Forestry*, vol. 92 (1994): 6–10. The

Brunson article also starts with the premise of little attention to wilderness in ecosystem management, but then turns to micro-management issues that can be learned from wilderness management such as horses versus llamas in wilderness, how much exotic weed control, and impact of people "recreating" in these areas.

19. I am just citing some of the better known and more visible people behind The Wildlands Project. See the 1992 special issue of *Wild Earth* from which this discussion draws, as well as subsequent issues that report and discuss ongoing analysis and particulars of the project.
20. This follows Reed Noss in his and Allen Y. Cooperrider's *Saving Nature's Legacy: Protecting and Restoring Biodiversity* (Washington, D.C.: Island Press, 1994). For a larger discussion of both The Wildland Project and the conservation biology principles that underlie it, this is the book to read.
21. Aldo Leopold, *A Sand County Almanac* (New York: Oxford University Press, 1948). For a broader discussion, see again Noss and Cooperrider.

## CHAPTER FOUR: What About Native Americans and Their Lands?

1. Chief Joseph, 1877, quoted in Peter Nabokow, ed., *Native American Testimony* (New York: Penguin Books, 1992).
2. The popular movie *Dances With Wolves* is an exception to these ethnocentric and biased views, though it erred on the other side by romanticizing the Indian culture and lifestyle, and was a story based not on actual events but the imagination of a screenwriter.
3. For a poignant discussion, see Charles F. Wilkinson, *The Eagle Bird: Mapping a New West* (New York: Pantheon Books, 1992).
4. This has been widely documented both in terms of overall trends and for specific groups. See, for example, Jack Chen, *The Chinese of America* (New York: Harper and Row, 1980); F. Cordasco and E. Bucchioni, *The Italians* (Clifton, N.J.: A.M. Kelley, 1974).
5. I say it is fairly recent because some of our best historians simply ignored the Indians. Frederick Jackson Turner was, and still is, considered by many to be the premier Western historian, but Jackson wrote little about Indians and what he did was racist in content. For less ethnocentric treatments of this history, see Alvin M. Josephy, Jr., *The Nez Perce Indians and the Opening of the Northwest* (New Haven: Yale University Press, 1965); Vine DeLoria, Jr., and Clifford M. Lyte, *The Nations Within: The Past and Present of Indian Sovereignty* (New York: Pantheon Books, 1984); A.M. Gibson, *The American Indian: Prehistory to Present* (Englewood Cliffs, N.J.: Prentice Hall, 1981).

6. See for example M.J. Bowden, "The Invention of American Tradition,": *Journal of Historical Geography*, vol. 18 (1992): 3–26.
7. See Karl Butzer, "The Indian Legacy in the American Landscape," in Michael P. Conzen, ed., *The Making of the American Landscape* (Boston: Unwin Hyman, 1990), pp. 27–50; Martin J. Bowden, "The Invention of American Tradition," *Journal of Historical Geography*, vol. 18 (1992): 3–26; Jack Weatherford, *Native Roots: How the Indians Enriched America* (New York: Crown Publishing, 1991).
8. William M. Denevan, "The Pristine Myth: The Landscape of America in 1492," *Annals of the Association of American Geographers*, vol. 82 (1992): 369–385; Martin W. Lewis, *Green Delusions: An Environmentalist Critique of Radical Environmentalism* (Durham: Duke University Press, 1992).
9. Peter G. Boag, *Environment and Experience: Settlement Culture in Nineteenth Century Oregon* (Berkeley: University of California Press, 1992).
10. I certainly don't mean to imply that this is universal. Living in place for thousands of years has led to conflicts and human atrocities in Europe and other places, most recently in the former Yugoslavia. Nor is nature necessarily a provider of good human values, as critics of deep ecology like to point out in the case of the nature-loving Hitler.
11. See Martin Bowden, note 6.
12. David Suzuki and Peter Knudtson, *Wisdom of the Elders: Honoring Native American Visions of Nature* (New York: Bantam Books, 1992).
13. Gundars Rudzitis and Jeffrey Schwartz, "The Plight of the Parklands," Environment, vol. 25 (1983): 6–11, 33–38.
14. Keith Petersen, *River of Life, Channel of Death* (Lewiston: Confluences Press, 1995).
15. The dams that were built with federal subsidies have yet to pay for themselves. The users—whether corporations, farmers, or consumers—receive a considerable subsidy still, relative to what other users pay private utilities. When the Clinton Administration proposed selling Washington Water Power Company to private owners to help decrease the federal budget deficit, the conservative congressmen and senators in the region reacted with horror, although normally they are proponents of the private market. The reason is obvious. Under private ownership and with no federal subsidies, it is estimated that electricity rates would increase by 20 percent or more.
16. In some sense, this is part of what is happening in post-Soviet Eastern Europe, Russia, and her former colonial states. They have been set adrift, and told to convert or drown. But, and it is a BIG But, Russia and former Soviet-dominated states and countries are receiving international aid from the World Bank and other organizations. The Asian "miracle" 'countries also received massive sums of international aid after

World War II. Cutting the tribes loose and expecting them to survive on their current land base, and to go out tribe by tribe to seek international aid would give them little chance for a bright future.

17. Quoted in Leonard A. Carlson, *Indians, Bureaucrats, and Land: The Dawes Act and the Decline of Indian Farming* (Westport: Greenwood Press, 1981).
18. Terry I. Anderson and Dean Leuck, "Land Tenure and Agricultural Productivity on Indian Reservations," *Journal of Law and Economics*, vol. 35 (1992): 427–454.
19. Charles Wilkinson, "Perspectives on Water and Energy in the American West and in Indian Country," South Dakota Law Review, vol. 26 (1981): 393–404. See also his 1987, *American Indians: Time and the Law* (New Haven: Yale University Press, 1987).
20. Robert Sassaman and Robert Miller, "Native American Foresty: Native Americans and the Bureau of Indian Affairs Are Cooperatively Managing Tribal Forest Lands," *Journal of Forestry*, vol. 84 (1986): 26–31.
21. Quoted in Peter C. Maxfield, Mary Francis Dieterich, and Frank J. Trelease, *Natural Resources Law on American Indian Lands*, Rocky Mountain Mineral Law Foundation, 1977.
22. Wendell Berry, *Another Turn of the Crank*, Counterpoint Press, 1995. See also Marshall Pecore, "Menominee Sustained-Yield Management: A Successful Land Ethic in Practice," *Journal of Forestry*, vol. 90 (1992): 12–16. Such practices, of course, are not limited to just Indians. For an example of how private non-Indians carry out similar practices, see Tim Foss, ft. 10, Chapter 3. Europe as a whole does not provide a good example, where sustaining forests for 140 years or more has not occurred.
23. L.C. Rule, "The Winnebago Tribe's Land-Use Planning: Alternatives in Forestry and Agriculture," *Journal of Forestry* , vol. 84 (1995): 26–31.
24. Robert Williams Jr., *The American Indian in Western Legal Thought* (New Haven: Yale University Press, 1990).
25. Ted Strong, "Tribal Rights to Fish," *Journal of Forestry* , vol. 92 (1994): 34.

## CHAPTER FIVE: Why Not Sell Off America's Wildlands?

1. Dramatic documentation of such corporate policies is provided by award-winning journalist Richard Manning in his *Last Stand: Logging, Journalism, and the Case for Humility.* (Salt Lake City: Peregrine Smith Books, 1991).
2. Michael D. Bowes and John V. Krutilla, *Multiple-Use Management: The Economics of Public Forestlands* (Washington, D.C: Resources for the Future and John Hopkins University Press, 1989).

3. The National Environmental Policy Act of 1969, the basis for the preparation of all those Environmental Impact Statements the federal land management agencies have done, requires that they develop better ways of estimating non-economic values.
4. For example, see Donald McCloskey, *Knowledge and Persuasion in Economics* (Cambridge, MA: Cambridge University Press, 1994); Paul Krugman, *Development, Geography, and Economic Theory* (Cambridge, MA: The MIT Press, 1995); Jack L. Knetsch, "The Changing Natural Resource Economy: Towards a Better Understanding," in Carmi Weingrod, ed., *Nature Has No Borders . . .* (Washington, D.C.: NPCA, 1994).
5. My discussion here owes much in particular to discussions with and the work of Jack Knetch and Tom Power.
6. John E. Keith, Christopher Fawson, and Van Johnson, "Wilderness Designation in Utah: Urban and Rural Willingness to Pay," Paper presented at Western Regional Science Meetings, Napa, California, February 1996.
7. C.A. Pope, III, and J.W. Jones, "Value of Wilderness Designation in Utah," *Journal of Environmental Management,* vol. 30 (1990): 157–174; R.G. Walsh, J.B. Loomis, and R.A. Gillman, 1984, "Valuing option, existence, and bequest demands for wilderness," *Land Economics,* vol. 60 (1984): 14–29.
8. Of course, a loss to one person may be a gain to another, and vice-versa. While logging and other activities may create losses of wilderness to many, the scaling back of timber harvesting to conserve or prevent loss of wilderness may create a sense of loss to loggers, especially if they lose their jobs.
9. This quote is attributed to John Maynard Keynes, who also said that, because of this short-run focus on maximizing profits, we must pretend that what is foul is fair, and what is fair is foul.
10. R.H. Thaler, "Some Empirical Evidence on Dynamic Inconsistency," *Economic Letter,* vol. 8 (1981): 201–207; G. Lowenstein and D. Prelec, "Anomalies in Intertemporal Choice: Evidence and an Interpretation," *The Quarterly Journal of Economics,* vol. 108 (1992): 573–597.
11. Having grown up in the New York City area, finding an auto mechanic I could trust and who proved worthy of that trust was a major source of satisfaction. After many lessons in the mechanics' school of hard knocks, I was lucky enough, after being sworn to secrecy, to find through word of mouth an honest mechanic. He was constantly overworked, but was much beloved by all his customers. Whenever I would bring in my treasures, a 1949 DeSoto or later my 1955 or 1961 Chevy, I would buy coffee and doughnuts from the diner across the street for Sal and whoever his current helper-apprentice happened to be.

12. D. Kahneman, J.L. Knetch and R.H. Thaler, "Fairness as a Constraint on Profit Seeking: Entitlements in the Market," *American Economic Review,* vol. 76 (1986): 728–741.
13. D.W. Meinig, "Spokane and the Inland Empire: Historical Geographical Systems and a Sense of Place," in D.H. Stratton, ed., *Spokane and the Inland Empire: An Interior Northwest Anthology* (Pullman, WA: Washington State University Press, 1991), pp. 1–32.
14. J.L. Knetch, 1990, "Environmental Policy Implications of Disparities Between Willingness to Pay and Compensation Demanded Measures of Value," *Journal of Environmental Economics and Management,* vol. 18 (1990): 227–206.
15. This discussion follows H. Rolston, III, "Valuing Wildlands," *Environmental Ethics,* vol. 14 (1985): 23–48 and M. Sagoff *The Economy of the Earth* (Cambridge: Cambridge University Press, 1988).

## CHAPTER SIX: How Does the American Public Want Wilderness Managed?

1. Riley E. Dunlap, "Polls, Pollution, and Politics Revisited: Public Opinion on the Environment in the Regan Era," *Environment,* vol. 26 (1987): 6–11, 32–37; see also his "Trends in Public Opinion Toward Environmental Issues: 1965–1990," *Society and Natural Resources,* vol. 4 (1991): 285–312; and Riley E. Dunlap and Rik Scarce, "The Polls-Poll Trends, Environmental Problems and Protection," *Public Opinion Quarterly,* vol. 55 (1991): 713–734.
2. R.C. Mitchell, "Public Opinion and Environmental Policies in the 1970s and 1980s," in N.J. Vig and M.E. Krafts, eds., *Environmental Policies in the 1980s: Reagan's New Agenda* (Washington, D.C.: Congressional Quarterly Press, 1984), pp. 51–74.
3. R Inglehart, *Culture Shift in Advanced Industrial Society* (Princeton: Princeton University Press, 1990).
4. L. Harris, V.L. Tarrance, and C.C. Lake, "The Rising Tide: Public Opinion, Policy and Politics," Americans for the Environment, Washington, D.C., 1989.
5. Anonymous, "Support for Land Stewardship Remains Strong Despite Economy," Montana's Tomorrow, 1–2.
6. J. Utter, "Opinions of Montanans on Wilderness and Resource Development," *Journal of Forestry,* vol. 81 (1989): 435–437.
7. Gundars Rudzitis and Harley E. Johansen, "How Important Is Wilderness? Results from a United States Survey," *Environmental Management,* vol. 15 (1991): 227–233.

8. Bruce Shindler, Peter List, and Brent S. Steel, "Managing Federal Forests: Public Attitudes in Oregon and Nationwide," *Journal of Forestry*, July 1993, 36–42.
9. G. Rudzitis, C. Watrous, H.E. Johansen, "Attitudinal Survey of Upper Columbia Residents," Working Paper, *Migration Regional Development and the American West Project*, Department of Geography, University of Idaho, 1995.
10. S. Elway, "Washington's Environmental Movement Is Alive and Well," *The Elway Poll*, September 1993, 1–6.
11. See footnote 9.
12. For example, in a survey in Chelan County, Washington, only 18 percent considered themselves to be politically liberal. See L. Krull, *Chelan County and the "New" American West*, MS thesis, Department of Geography, University of Idaho, 1985. Chelan County also was included in the survey results cited in footnote 9.
13. J.C. Archer and F.M. Shelley, *American Electorial Mosaics* (Washington, D.C.: AAG, 1986).
14. Nancy Langston, *Forest Dreams, Forest Nightmares: The Paradox of Old Growth in the Inland West* (Seattle: University of Washington Press, 1995). Foresters who wanted to save forests instead destroyed them. She dramatically shows that, today, similar studies are being used to justify the very actions that created current ecological problems. Somewhat incredibly, she shows how scientists recognize this and then proceed to recommend actions that will only make the current situation worse. Grey science at its worst! This raises the question again, "For whom are the public lands being managed?"
15. For a good overview of how this works, see, R. O'Toole, *Reforming the Forest Service* (Washington, D.C.: Island Press, 1988).

## CHAPTER SEVEN: Wilderness and the Communities of the American West

1. W. Stegner, *The American West as Living Space* (Ann Arbor: University of Michigan Press, 1990).
2. D.E. Popper and F.J. Popper, 1987, "The Great Plains: From Dust to Dust," *Planning*, vol. 53 (1987): 12–18.
3. G. Rudzitis and H. E. Johansen, *Amenities, Migration and Nonmetropolitan Regional Development*, Report to the National Science Foundation, June 1989. See also G. Rudzitis and H.E. Johansen, "Migration into Western Wilderness Counties: Causes and Consequences," *Western Wildlands*, Spring 1989: 19–23.

4. For more details, see P. Hauser, "The Census of 1980," *Scientific American*, vol. 245 (1981): 53–61.
5. A good historical geography of cities in the modern American West is C. Abbott, *The Metropolitan Frontier* (Tucson: University of Arizona Press, 1993).
6. Some good discussions can be found in B.J.L. Berry, "The Counter-Urbanization Processes: How General?" in N.M. Hansen, ed., *Human Settlement Systems* (Cambridge: Ballinger Publishing, 1978), pp. 25–49; R.L. Morrill, "Stages in Patterns of Population Concentration and Dispersion," *Professional Geographer* (February 1979): 55–65; J.M. Wardwell and D.L. Brown, "Population Redistribution in the United States During the 1970s," in D.L. Brown and J.M. Wardwell, eds., *New Directions in Urban-Rural Migration* (New York: Academic Press, 1980), pp. 137–162.
7. D.J. Morgan, *Patterns of Population Distribution: A Residential Preference Model and Its Dynamic*, Research Paper, No. 176, Department of Geography, University of Chicago, 1978; D.A. Dillman, "Residential Preferences, Quality of Life and the Population Turnaround," *Agricultural Journal of Economics*, vol. 61 (1979): 960– 966.
8. See J.B. Lansing and E. Mueller, *The Geographical Mobility of Labor* (Ann Arbor: Survey Research Center, University of Michigan, 1967); H.E. Johansen and G.V. Fuguitt, *The Changing Rural Village: Demographic and Economic Trends Since 1950* (Cambridge: Ballinger Publishing, 1984).
9. See the now classic work of M. White and L. White, *The Intellectual Versus the City* (Cambridge: Harvard University Press, 1962); and Y. Tuan, "Strangers and Strangeness," *Geographical Review*, vol. 76 (1986): 10–19.
10. For example, J.D. Williams and A.J. Sofranko, "Motivation for the In-Migration Component of Population Turnaround in Nonmetropolitan Areas," *Demography*, vol. 16 (1979): 235–239; L.H. Long and D. DeAre, *Migration to Nonmetropolitan Areas: Appraising the Trend and Reasons for Moving*, Special Demographic Analysis, CDS-80-2, U.S. Bureau of the Census, Washington, D.C., 1980.
11. Dean Judson, "Survey of Migrants to Oregon," unpublished paper, Center for Economic Development, University of Nevada, Reno, 1995; G. Rudzitis, C. Watrous and H.E. Johansen, "Attitudinal Survey of Upper Columbia River Residents," Department of Geography Discussion Paper, University of Idaho, 1995.
12. The discussion in this section is based on my participating in a workshop on the "interface" issues and presenting a paper there. See G. Rudzitis and H.E. Johansen, "Why Do People Move to Wildland Areas?" unpublished paper prepared for Workshop on Urban Wildland Issues and Opportunities, Grey Towers National Historic Landmark and Pinchot Institute for Conservation Foundation, 1988.

13. See W. Riebsame, H. Gosnell, and D. Theobald, "Land Use and Cover Change in the U.S. Rocky Mountains I: Theory, Scale and Pattern," *Mountain Research and Development*, forthcoming, 1996; W. Wykoff and K. Hansen, "Settlement, Livestock Grazing, and Environmental Change in Southwest Montana," *Environmental History Review*, vol. 15 (1991): 45–71.
14. The same slogan, with the names of Montana, Wyoming, or other states substituted, can be seen in shops throughout the West.
15. I recall that when I taught at the University of Texas in Austin and told a colleague that I was going to interview for a job at the University of Idaho, he replied "They don't pay well, or didn't when I interviewed a number of years back, but they told me, like other faculty I could supplement my salary by hunting and saving money on the meat I had to buy." Another colleague who had taught in Montana was horrified, saying, "You'll hate it. People go hunting up there, and classes empty out during hunting season. It's no place for a cosmopolitan New Yorker." Fortunately they were both wrong.
16. For example, in Idaho the maximum penalty for carelessly killing someone, and normally offering the "I thought he was an elk" defense, is a five year loss of your hunting license.

## CHAPTER EIGHT: Wilderness and Economies of the Old and New West

1. The debate took place within two articles, a reply, and a rejoinder. D.C. North, "Location Theory and Regional Economic Growth," *Journal of Political Economy*, vol. 63 (1955): 243–258; C.M. Tiebout, "Exports and Regional Economic Growth," *Journal of Political Economy*, vol. 64 (1956): 160–169.
2. Interestingly, North used the example of the cut-over region in the Great Lakes area as an example of stranding a region by cutting down all the trees.
3. The first, to no one's surprise, was Harvard, with the wealth coming largely from alumni.
4. With no malice or criticism intended, see Michael P. Malone and Richard W. Etulain, *The American West* (Lincoln: University of Nebraska Press, 1989).
5. The theoretical argument for the critical role of the entrepreneur is based on the work of economist J.A. Schumpeter, *A Theory of Economic Development* (Cambridge: Harvard University Press, 1934).
6. E. Malecki, "New Firm Startups: Key to Rural Growth," *Rural Development Perspectives*, vol. 4 (1988): 18–23.

7. T.M. Power, *Extraction and the Environment: The Economic Battle to Control Our Natural Landscapes* (Washington, D.C.: Island Press. 1995); R. Rasker, "A New Look at Old Vistas: The Economic Role of Environmental Quality in Western Public Lands," *University of Colorado Law Review,* vol. 65 (1994): 369–399.
8. L.C. Krull, *Chelan County and the "New" American West,* MS thesis, Department of Geography, University of Idaho, 1995.
9. W.B. Beyers, "Trends in Service Employment in Pacific Northwest Counties: 1974–1986," *Growth and Change,* Fall 1991, pp. 28–50.
10. G. Rudzitis and H.E. Johansen, "How Important Is Wilderness? Results of a United States Survey," *Environmental Management,* vol. 15 (1991): 227–233.
11. P.E. Graves and P. Linneman, "Household Migration: Theoretical and Empirical Results," *Journal of Urban Economics,* vol. 6 (1979): 383–404.
12. B. Marsh, "Continuity and Decline in the Anthracite Towns of Pennsylvania," *Annals, Association of American Geographers,* vol. 77 (1987): 337–352.
13. S. Weil, *The Need for Roots* (New York: Harper and Row, 1952).
14. W. Berry, *Home Economics: Fourteen Essays* (San Francisco: North Point Press, 1987).
15. E. Relph, *Place and Placelessness* (Toronto: University of Toronto Press, 1986).
16. D. Tall, *From Where We Stand: Recovering a Sense of Place* (New York: Alfred Knopf Publishers, 1993).
17. The Mideast and other OPEC oil companies are the exception to the rule. A dependence on extraction of natural resources for most countries of the world has put them in a vulnerable position, and with little control over prices.
18. G. Spence, *With Justice for None Destroying an American Myth* (New York: Times Books, 1989).
19. S. Weil, *The Need for Roots* (New York: Harper & Row, 1952).
20. One private timber company manager, when asked how they could log a whole mountain, laughed and said it was easy. He said, if needed, they could probably log all of the public lands within a year, leaving little standing.
21. Richard Manning, *Last Stand, Logging, Journalism, and the Case for Humility* (Salt Lake City: Gibbs-Smith, 1991).
22. J.D. Johnson and R. Rasker, "Local Government: Local Business Climate and Quality of Life," *Montana Policy Review,* Fall 1993, pp. 11–19; R. Rasker and D. Glick, "Footloose Entrepreneurs: Pioneers of the New West?" *Illahee,* vol. 10 (1994): 34–43.; W.B. Beyers, D.P. Lindahl, and E. Hamill, "Lone Eagles and Other High Flyers in the Rural Producer Ser-

vices," presented at Pacific Northwest Regional Economic Conference, May 1995, Missoula, MT; J.D. Johnson and R. Rasker, "The Role of Economic and Quality of Life Values in Rural Business Locations," *Journal of Rural Studies*, vol. 11 (1995): 405–416.

23. For example, in 1977 when I went to teach at the University of Texas, an environmental economics course was not part of the curriculum in the Economics Department. The same was true in 1983 when I came to the University of Idaho. In the spirit of full disclosure, I should point out that, in graduate school, I worked as a research assistant for George Tolley.
24. I put deviant in quotes because anthropologists and sociologists argue that what is deviant in one society may not be in another, excluding obviously violent and despicable acts. Even this creates problems for people who cite Greek philosophers as exemplary examples of a golden age, but cringe or explain away the slavery, sexual abuse of women, or taking of young boys as lovers by these very same people.
25. For a good economic discussion of sense of place, see R. Bolton, "Place Prosperity vs. Place Prosperity Revisted: An Old Issue with a New Angle," *Urban Studies*, vol. 29 (1992): 185–203.

## CHAPTER NINE: "It's My West, Not Yours"

1. Erik Larson, "Unrest in the West," *Time*, October 23, 1995, pp. 52–66. The *Time* cover was entitled "Don't Tread on Me: An Inside Look at the West's Growing Rebellion." There have been any number of newspaper and journal articles around this theme.
2. Calling out federal troops was not an unusual practice in the West. When miners and others went out on strike, for example, federal troops often were called in to quell the "uprising".
3. Comments of Robert Burford, Director of the Bureau of Land Management, quoted in Gundars Rudzitis, "Federal Lands: Wilderness Management Policy," *Environment*, May 1984, pp. 2–4.
4. For more on the Sagebrush Rebellions, see William L. Graf, *Wilderness Preservation and the Sagebrush Rebellions* (Savage, Md.: Rowman & Littlefield, 1990); R. McGreggor Cawley, *Federal Land, Western Anger: The Sagebrush Rebellion and Environmental Politics* (Lawrence: University of Kansas Press, 1993).
5. Most of what has been written about the Wise Use Movement has been either in newspapers or in articles in the journals of environmental organizations, which obviously give it their own slant. Some of these include: Paul Rauber, "Wishful Thinking: Wise Use Cowboys Try to Rewrite the Constitution," *Sierra*, vol. 79 (1994): 39–42; Robert Henel-

ly, "Getting Wise to the Wise Guys: The New Sagebrush Rebels Ride East", *Amicus Journal,* vol. 14 (1992); Richard Stapleton, "Greed vs. Green", *National Parks,* vol. 66 (1992): 32–37; James Ridgeway and Jeffrey St. Clair, "This Land is Our Land: Inside the Property Rights Movement," *Wild Forest Review,* vol. 2 (1995): 14–20. For a book-length treatment of the movement, see David Helvarg, *The War Against the Greens: The Wise Use Movement, the New Right, and Anti-Environmental Violence* (San Francisco: Sierra Books, 1994). I also benefited from Chris Wall, "The Wise Use Movement: Who Are These Guys and What Do They Want?" unpublished paper, 1994.

6. As an aside, many people probably don't realize that a tenured position at a college or university does not mean that you have a job for life. In most places, financial exigency or downsizing can be reasons to give you a 90 day notice to find another job, though one years' notice is much more likely. All things considered, the tenured professor still has much more security than the average timber worker or executive, for that matter.
7. Indeed, the local paper where I live periodically publishes letters complaining about the treatment of long-term workers at the timber mill about 30 miles from where I live. A typical letter might be about someone being fired after 20 years with few benefits, or even more poignantly about a worker who was seriously injured or killed at the mill and the insensitivity of the management in taking any responsibility. However accurate and truly representative such letters are, they do indicate the callousness with which workers and their survivors feel they are being treated by their resource employers as we head into the twenty-first century.
8. Actually, to say any losses is not correct. Normally such compensation would be required when actual property values were estimated to fall by a certain percentage, such as 10 or 20 percent.
9. I recall one person giving me this as the reason why he moved out from the city. Later, by coincidence, I was talking to his rural neighbor who had lived in the area all his life and who was now complaining about this newcomer who had fenced off the road running through his property, restricted access, and posted no hunting signs even though he hunted himself. "It was not a neighborly thing to do. My boys and I had been walking and hunting those lands for years. And he got nasty about it," complained the old-timer. Fleeing restrictions, this former urbanite set his own and disrupted local access traditions, setting in motion long-lasting hard feelings.
10. Eric T. Freyfogle, "Land Ownership, Private and Wild," *Wild Earth,* vol. 5 (1996): 71–77. For more on the role of private property rights in United

States history, see Eric T. Freyfogle, *Yale Law Journal*, vol. 94 (1985): Richard E. Foglesong, *Planning the Capitalist City: The Colonial Era to the 1920s* (Princeton: Princeton University Press, 1986); Sam Bass Warner, *The Urban Wilderness: A History of the American City* (New York: Harper & Row, 1972).

11. Olen Paul Matthews, "The Supreme Court, the Commerce Clause, and Natural Resources," *Environmental Management*, vol. 12 (1988): 413–427.
12. Philip Weiss, "Enter Government Hating, Home Schooling, Scripture Quoting Idaho, the New Leave- Me-Alone America at Its Most Extreme off the Grid," *New York Times Magazine*, January 8, 1995, pp. 24–34, 48–52. The impact of the article was such that several of my students from outside Idaho said that their parents called after reading the article to ask if they were safe living in Idaho.
13. Paula Young, *Analysis of Wilderness County Migration Flows: 1975–1980*, Master's thesis, University of Idaho, 1988.
14. The all-white characteristics exclude the American Indians living in the area because they live on their reservations and interact hardly at all on a day-to-day basis with the surrounding white culture.
15. See, for example, Robert H. Blank, *Individualism in Idaho: The Territorial Foundations* (Pullman: Washington State University Press, 1988) and Carlos Schwantes, *The Pacific Northwest: An Interpretive History* (Lincoln: University of Nebraska Press, 1989). The history of Idaho and the inner West also includes sites for Japanese "detention" or "concentration" camps during World War II, a shameful national paranoia that did not extend to white Germans in the United States.
16. David Briggs, "United Methodist Church Seeks to Combat Rural Racism," Associated Press reported in *Lewiston Tribune*, May 11, 1996, 13A).
17. Arthur M. Schlesinger, Jr., *The Disuniting of America* (New York: W.W. Norton & Company, 1992).
18. The statistics cited in this discussion are taken from a report, *False Patriots: The Threat of Antigovernment Extremists*, prepared by the Militia Task Force, Joseph T. Roy Sr., Director, of the Southern Poverty Law Center, Montgomery, AL, 1996; For some book-length studies of parts of the movement and their causes, see James A. Aho, *The Politics of Righteousness: Idaho Christian Patriotism* (Seattle: The University of Washington Press, 1990); James Williams Gibson, *Warrior Dreams: Paramilitary Culture in Post-Vietnam America* (New York: Hill and Wang, 1994); Morris Dees and James Corcoran, *Gathering Storm: America's Militia Network* (New York: HarperCollins, 1996). For a somewhat lighter perspective, see Brent Israelsen, "The Wacky West? Montana Looks Weird to Rest of U.S.," *Salt Lake Tribune*, April 15, 1996.

19. Anonymous, "Militia Launch Counter-Intelligence Campaign," *SPLC Report*, December 1995, p. 1. The Aryan Nations Congress had about 200 attendees.
20. As one example, the Wilderness Society in 1995 decided to put more of its efforts into providing assistance to formerly resource-based communities. Personal conversation with Ray Rasker, The Wilderness Society, October 1995.
21. Gundars Rudzitis, Christy Watrous, and Harley Johansen, *Public Views on Public Lands: A Survey of Interior Columbia River Basin Residents*, Working Paper of The Migration, Regional Development, and Changing American West Project, Department of Geography, University of Idaho, November 1995.
22. Wallace Stegner, *The American West as Living Space* (Ann Arbor: University of Michigan Press, 1987).
23. More information about the North International Cascade Park proposal and how it fits in with other changes taking place in the region can be gotten from: Carmi Weingrod, ed., *Nature Has No Borders* (Washington, D.C.: NPCA, 1994); Mitch Friedman and Paul Lindholdt, eds., *Cascadia Wild: Protecting an International Ecosystem* (Bellingham: Greater Ecosystem Alliance, 1993).
24. There have been a large number of studies done by the agencies as well as by academics on the impact of people on the natural environment in our national forests. How many people disturb trails, campgrounds, and the natural surroundings is an area not lacking in studies.

## CHAPTER TEN: Future Directions for Wilderness

1. Kenneth Arrow, *Social Choice and Individual Values* (New Haven: Yale University Press, 1963).
2. Although there has been increased discussion about the need to train foresters and other resource management majors in undergraduate colleges, this seems a commendable, but impractical approach. For example, a review of requirements for a major in forest resources or forest products shows that there are very few required courses in the biological sciences, social sciences, or environmental philosophy.
3. Michael Frome, *The Battle for the Wilderness* (New York: Praeger Publishing, 1974). I also pointed out the need to consider options to the current management of wilderness by four different agencies in my article, "Federal Lands: Wilderness Management Policy," *Environment*, May 1984, pp. 2–4.
4. See Michael Frome, *Regreening the National Parks* (Tucson: University of Arizona Press, 1992).

5. The charges were primarily for influence peddling and lying, after he left the Interior Department, not for activities that took place while he was in office. Nonetheless, it does taint his reputation and that of the Department of the Interior, where scandals and cover-ups are a part of its history.
6. Charles Wilkinson has written in a variety of places about the impact of these laws. A cogent summary, followed by chapters exploring each law in turn, can be found in his *Crossing the Next Meridian: Land, Water, and the Future of the West* (Washington, D.C.: Island Press, 1992).
7. I say "normally" because I do realize that individual congresswomen and senators often do intervene in individual cases, or try to get waivers for individuals, companies, or corporations from environmental regulations, for example, but this is not the day-to-day practice.
8. Richard White, "Indian Land Use and the National Forests", in H.K. Steen, ed., The *Origin of the National Forests* (Durham: Forest History Society, 1992); Stephen Pyne, *Fire in America: A Cultural History of Wildland and Rural Fire* (Princeton: Princeton University Press, 1982); Stephen W. Barrett and Stephen F. Arno, "Indian Fires as an Ecological Influence in the Northern Rockies," *Journal of Forestry*, October 1982, pp. 647–650.
9. See Richard White, note 8; "Addressing the Cultural Resources of the Northern Cascades", a series of three articles with the same general title, "1: From the Upper Skagit Tribe" by Doreen M. Malony; 2: "From the Sto:lo Nation" by Larry Commodore; 3: "From the Nlaka" Pamux Nation" by Chief Bob Pasco, in Carmi Weingrod, ed., *Nature Has No Boundaries* (Washington: D.C.: NPCA, 1994); Vine De Loria, *God Is Red: A Native View of Religion* (Golden, CO: Fulcrum Publishing, 1994).
10. I am referring not to specific churches, but actual places of importance to a religion. Those are in Israel for Jews and Christians, and yet, unlike Moslems, very few Jews or Christians ever make a pilgrimage to Israel to visit places considered holy by their religion. I also am ignoring the sacred places and practices of some New Age groups that attribute sacredness to particular places in the American West, such as in New Mexico. Unfortunately, some of these groups have tried to graft Native American practices onto theirs in a very superficial way.
11. Doreen Maloney, personal communication, March 26, 1994.
12. In states such as Idaho we have checkoffs on our state income tax form to designate how much of our taxes we want to contribute to the Fish and Wildlife Fund.
13. Richard Morrill, "Local and Regional Population Growth: The Limits to Intervention," *Northwest Environmental Journal*, vol. 5 (1989): 274–298.
14. For some examples, see John McPhee, *The Control of Nature* (New York: Farrar Strauss & Giroux, 1989).

# Index